Accessible Mathematics Education for Students With Visual Impairments

Derrick W. Smith, Editor

APHPress
American Printing House for the Blind

Library of Congress Cataloging-in-Publication Data

Names: Smith, Derrick W., 1977- editor
Title: Accessible mathematics education for students with visual impairments / Derrick W. Smith, editor.
Description: Louisville, KY : APH Press, [2026] | Series: APH press stem | Includes bibliographical references and index.
Identifiers: LCCN 2025049886 (print) | LCCN 2025049887 (ebook) | ISBN 9781950723263 paperback | ISBN 9781950723270 ebook
Subjects: LCSH: Mathematics--Study and teaching | Blind--Education | People with visual disabilities--Education | AMS: Mathematics education -- Education and instruction in mathematics -- Teaching methods and classroom techniques
Classification: LCC HV1664.A7 A23 2026 (print) | LCC HV1664.A7 (ebook)
LC record available at https://lccn.loc.gov/2025049886
LC ebook record available at https://lccn.loc.gov/2025049887

This book is dedicated to Donte, Jason, Jennifer, April, Jennifer, Jamie, and Dustin, my first class of students who helped me fall in love with teaching.

CONTRIBUTORS

EDITOR

Dr. Derrick Smith is a Professor of special education and Dean of the College of Education at Auburn University at Montgomery. His research primarily focuses on STEM education for students with visual impairments, particularly how technology can support mathematics education. He has been awarded and collaborated on multiple research projects funded by the National Science Foundation (NSF) and the U.S. Department of Education. He has published and presented on various topics related to STEM education for students with visual impairments both nationally and internationally.

CHAPTER AUTHORS

Dr. Aina Appova is an Associate Professor of Mathematics Education at The Ohio State University. She earned her PhD in mathematics education from the University of Missouri–Columbia and holds a master's degree in mathematics from Central Michigan University. Dr. Appova is a fellow of the Center for the Study of Mathematics Curriculum (CSMC), a program supported by the National Science Foundation (Award No. ESI-0333879). Over the past decade, she has developed and taught mathematics content and methods courses for prospective PK–8 teachers. Her research focuses on the learning trajectories, professional capacity, development, and continuum of both preservice and inservice PK–8 mathematics teachers and teacher educators. Recently, as Principal Investigator on a funded grant project, Dr. Appova led an initiative that (statistically) significantly enhanced children's mathematics learning and achievement within a high-need, large urban school district. This work involved a detailed examination of the pedagogical content knowledge, instructional practices, and collaborative efforts of K–5 mathematics teachers and teacher educators.

Dr. Amanda Lannan is a Clinical Assistant Professor in the Visual Impairment Program at the University of Kentucky, where she works to increase equitable access to high-quality STEM education for students with visual impairments. Her research draws from disability studies in education and universal design for learning, focusing on expanding advanced mathematical opportunities for culturally and neurodiverse learners. She collaborates nationally through efforts such as serving on the DIAGRAM Center's Research Working Group, contributing to research on the impact of tactile graphics on STEM learning. A 2021 NSF CADRE Fellow, Dr. Lannan has a rich background as a classroom teacher across general, special, and visual impairment education. As a blind, disabled educator and researcher, she brings both lived experience and academic expertise to her advocacy for inclusive education. She challenges educators to take bold action and recognize students with disabilities as capable learners who deserve access, opportunity, and high expectations.

Dr. Sarah Roller Dyess is an Associate Professor in the Department of Curriculum and Instruction at the University of Alabama in Huntsville and a former middle school teacher. She teaches mathematics methods courses for secondary and elementary education majors, as well as mathematics content courses for elementary teachers. Her research focuses on teacher preparation, recruitment, and research-based strategies for effective mathematics instruction.

Dr. Sean Tikkun is an Assistant Professor at North Carolina Central University in the Visual Impairments Training Program. He holds a PhD in Educational Psychology, an M.S. in Education Media Design and Technology, an M.S. in Special Education, and a B.A. in Mathematics. Within his profession he serves as Treasurer of the Association for Education and Rehabilitation of the Blind and Visually Impaired (AERBVI). Illinois AERBVI awarded him the distinguished service award for his work in training assistive technology professionals for the Academy for Certification of Vision Rehabilitation and Education Professionals (ACVREP) and developing the first automated grading tool for braille. He is a member of the board of trustees for the Governor Morehead School for the Blind. His research focuses on cognitive factors for tactile learning and emerging technology applied to visual impairments.

Dr. Tiffany Wild is an Associate Professor in the Inclusive STEAM education section of the Department of Teaching and Learning at The Ohio State University. Her research focuses on making STEM content accessible for students with visual impairments. Her commitment to accessible STEM for students with visual impairments has been recognized worldwide through numerous presentations and publications.

INTRODUCTION

Derrick W. Smith

This book is really about *love*.

Now, most people would think that I might start this introduction by discussing my love for *mathematics* or talking about how this book is a *labor* of love. While those things are interesting, they are not what drove me to want this book to exist.

No, the truth is that I started writing this book in my heart and mind years ago, when I became the secondary mathematics teacher at the Alabama School for the Blind (ASB). In the summer of 2000, the school's math teacher abruptly took an administrative job at the Alabama School for the Deaf (ASD) a few weeks before the fall semester started. Through a twisting series of events, I was contacted by the principal of ASB to see whether I had any interest in helping them out by coming to teach for a year. The principal knew that I had just graduated with a mathematics degree and that I was unhappy in my job as a network administrator. I had never considered becoming a teacher, but when he contacted me out of pure desperation, I was intrigued by the opportunity. At the time, I was earning $20,000 a year, and this teaching job paid $29,760. Upon learning that information, I accepted the invitation to interview. For some reason—and despite me having relatively little background in teaching—they hired me.

After one week of observing other teachers working with these students at ASB, I was thrust into my classroom with little background in teaching. I spent a great deal of time reading books, researching "how to teach," and trying to develop my teaching philosophy and style. I quickly learned that I was not equipped to be a math teacher, much less a specialized teacher for students with visual impairments (TSVI). But I had a strong support team at ASB, including wonderful paraprofessionals who knew braille, and we trudged through the year together. I must admit that I did a horrible job that first year, but something changed within me: I fell in love!

I didn't fall in love with mathematics. I already appreciated mathematics.

Nor did I fall in love with teaching (at least not immediately). Though I quickly developed a strong respect for the artistry and science of this noble profession, I did not "love" it yet. That would come later, after I had spent more time doing it.

What I fell in love with was my students. Interacting with these teenagers every day was a true joy. I loved being part of their journey into adulthood. We laughed together and even cried together. As I taught math, some days were very hard because I was teaching content that many did not enjoy learning. But seeing their excitement, the light in their eyes when they "got it," made it all worthwhile. I quickly realized that these students with visual impairments did not allow their vision challenges to define them. They were just kids who wanted opportunities in life. I found great joy in providing that for them.

It was due to this love that, during my first year of teaching, I started working on my teaching certification and master's degree in special education. This love motivated me to start consuming everything I could about teaching math to students with visual impairments; I started learning not only literary braille but also Nemeth code almost immediately. When given the opportunity to return for a second year, the decision was easy: I did, and it was because of my students. I came to teaching—admittedly through the "back door"—for money, but I stayed because of love.

During my tenure as the secondary mathematics teacher at ASB and while I completed my master's degree in special education (with a specialty in collaborative teaching and visual impairments), I spent an extraordinary amount of time looking for strategies, interventions, tools, and technologies for teaching mathematics to students with visual impairments. While some work in this area had been completed, most notably by Gaylen Kapperman and Jodi Sticken, there was a tremendous gap between general education mathematics research and pedagogy and what was available to "my" students. I began creating modified tools and strategies to better support them—and while they were not always novel, they were novel to me. (As a teacher at ASB, I thought that I had invented a way to "speak math" while I wrote it on the board for my braille readers. Later, I learned that I was re-creating the efforts of others who had already devised different specifications for speaking math that avoided confusion for students with limited or no vision.) It was during my time working with my students that I determined that my career goal was to help students with visual impairments have appropriate—even equal—access to mathematics content.

That said, it is impossible for me to complete this introduction without discussing my affection for mathematics. Most teachers who teach math content (general educators, special educators, or TSVIs) are fighting against many factors that make our jobs challenging: Students sometimes struggle with this content and thus do not enjoy it. Students complain about not seeing the purpose of math in their future and often have adults in their lives who express similar ideas. Family members may say, "I'm not good at math," allowing these students to have an excuse not to be good at it either. For some students, the topic of math brings up dread, fear, and even hatred. Thus, the goal of the teacher of math is to ensure that these students' fears and dread are calmed by demonstrating for them the purpose of math in their lives and giving them a clear sense that understanding mathematics empowers them to understand so much more about the world around them.

Any teacher who teaches mathematics must have an attachment to the purpose of teaching math to others. In my opinion, math is critical for two primary reasons: communication and utility.

First, regarding communication, mathematics is a language. To be considered a language, a system of communication must have vocabulary, grammar, syntax, and people who use and understand it. Without going into deep details, mathematics meets all of these standards. Pure linguists may contend that math is not a language because it is primarily a written code rather than a spoken form of community. However, mathematics has long been considered "the language of science." I would take that a step further and say that math is the universal language of not only science but also medicine, engineering, business, finance, architecture, music, art, logic, and more. Having a solid foundation in mathematics gives our students the opportunity to excel in the field of their choosing, as they can literally "speak the language."

Secondly, mathematics is a utility—it is a tool with limitless use. Closely connected to the idea of math as a language, mathematics is often the tool that we use in order to not only communicate ideas but also to solve problems. Whether we are attempting to determine the tensile strength of a material, the best price to sell a product to achieve maximum profit, the measurements of a distant star, or the exact amounts of an ingredient to make the perfect cake, mathematics is often the best tool to make sense of the world and solve complex and real problems. Within mathematics as a tool is the concept of *numeracy*, which is the ability to understand and work with numbers. However, more than that, the ability to use numbers (i.e., data, often in numerical form) allows us to make informed decisions. When we have strong numeracy skills, we tend to make more informed decisions that lead to better outcomes.

Students with visual impairments and blindness are often at risk of not learning the mathematical skills necessary for them to communicate and solve problems that involve numbers. There are many factors that may impact their mathematics learning. Many aspects of mathematics content are highly visual: The content requires specialized notation, both in print and in braille; some concepts are commonly taught using spatial arrangements; aspects of mathematics require drawing; and they may require specialized tools (such as calculators, protractors, compasses, etc.), to name a few. Overcoming considerations like these requires the knowledge of both the TSVI and the general education mathematics teacher working in tandem to provide accessible instruction.

Mathematics is beautiful because it opens so many doors of opportunity. Students with visual impairments deserve the opportunity to learn and use mathematics in whatever way they decide. It is for this reason that this book was written. Each of the authors of this book has a passion to ensure that all students, including those with visual impairments, have the opportunity to excel in whatever passions they have in their lives. This book's authors include individuals with expertise in mathematics, mathematics education, and teaching students with visual impairments. All of them have experience teaching mathematics, and each chapter has at least one author who has taught mathematics to students with visual impairments.

This book is written primarily to support TSVIs, but I firmly believe that general education math teachers will benefit from its contents. The chapters are organized primarily by subcontent areas within mathematics. While there are a few chapters that focus solely on elementary grades, many of the chapters will address content from kindergarten to 12th grade. While we would love to answer every question a teacher may have within this text, we realize that is impossible. Therefore, we focus on addressing the areas where the impact associated with visual impairment would be the most pronounced when teaching and learning these concepts. Within these pages, we will not only discuss the challenges but also present strategies to best address those challenges.

In the end, everyone who contributed to this book wants its readers to feel better prepared to support "our" students. We want readers to leave this book with the tools they need to make a difference with every lesson and every math problem, bringing our students a rich, meaningful mastery of math as both a language to describe their world and a tool to understand it.

CHAPTER 1

MATHEMATICAL PRACTICES: HOW STUDENTS SHOULD ENGAGE IN LEARNING MATHEMATICS

Sarah Roller Dyess and Derrick W. Smith

INTRODUCTION TO THE STANDARDS OF MATHEMATICAL PRACTICE

Within the Common Core State Standards for Mathematics (NGA Center & CCSSO, 2010) are two types of standards: Standards for Mathematical Content and Standards for Mathematical Practice. The Standards for Mathematical Content describe the mathematical content students need to learn at each grade level, whereas the Standards for Mathematical Practice outline the "varieties of expertise that mathematics educators at all levels should seek to develop in their students. These [mathematical] practices rest on important 'processes and proficiencies' with longstanding importance in mathematics education" (NGA Center & CCSSO, 2010, p. 6). These bedrock "processes and proficiencies" include the National Council of Teachers of Mathematics' (NCTM, 2000) process standards, which include problem solving, reasoning and proof, communication, representation, and connections, as well as the five strands of mathematical proficiency outlined in the National Research Council/Mathematics Learning Study Committee report *Adding It Up* (2001), which include conceptual understanding, procedural fluency, strategic competence, adaptive reasoning, and productive disposition. Thus, the content standards identify *what* students need to learn, and the mathematical practices outline *how* the students should engage in learning that content (e.g., the mathematical skills they should be able to leverage in their practice).

As general educators work with teachers of students with visual impairments (TSVIs), it is important that TSVIs have a general understanding of not only *what* the math teacher is teaching but also *how* the students are engaging in the content. Understanding the purpose of the lesson (whether it focuses on knowledge or skills) and its underlying mathematical practices will help the TSVI determine the best way to collaborate with the general educator. For example, a specific lesson may focus more on utilizing one specific mathematical practice over another and thus may impact the lesson's goals and outcomes. This chapter introduces these mathematical practices so TSVIs can understand them and best support the students that we serve.

UNPACKING THE MATHEMATICAL PRACTICES

There are eight mathematical practices identified in the Common Core State Standards for Mathematics (see Sidebar 1.1), versions of which are also included in many state standards. Mathematical practices are not a checklist—they are various ways students can meaningfully engage in mathematics similar to the ways mathematicians do and use mathematics. For example, Mathematical Practice 1 (Make sense of problems and persevere in solving them) and Mathematical Practice 6 (Attend to precision) are habits of mind that students should regularly practice. Other mathematical practices, while not necessarily used daily, are nevertheless skills that students should utilize when applicable. The content and goals of a lesson will drive which mathematical practices teachers decide to have students leverage in the lesson.

Sidebar 1.1 Mathematical Practices

MP1: Make sense of problems and persevere in solving them.
MP2: Reason abstractly and quantitatively.
MP3: Construct viable arguments and critique the reasoning of others.
MP4: Model with mathematics.
MP5: Use appropriate tools strategically.
MP6: Attend to precision.
MP7: Look for and make use of structure.
MP8: Look for and express regularity in repeated reasoning.

Note. Mathematical Practices are from NGA Center & CCSSO, 2010.

To support readers in understanding the mathematical practices, this chapter broadly describes each individual mathematical practice, including the main ideas of each practice. The goal is to provide TSVIs with a basic overview that captures the heart of each mathematical practice.

MP1: Make Sense of Problems and Persevere in Solving Them

Mathematical Practice 1 (MP1) focuses on problem solving and contains two parts: (a) making sense of problems and (b) persevering in solving them. In terms of making sense of problems, students engaged in MP1 are able to approach open-ended problems with no designated solution path and use their problem-solving skills to make sense of what the problem is asking, decide on tools or strategies to use to solve the problem, find an answer, and then go back and check to make sure the answer makes sense. The second part of MP1 is persevering in problem solving, which focuses on students' abilities to persist, including making changes when strategies do not work or when an answer turns out to be incorrect.

To support students in making sense of problems and persevering in solving them, teachers may want to consider the four steps in Pólya's problem-solving process (Pólya, 1945) and think about ways they can provide guidance at each stage.

- **Step 1:** Make sense of the problem. When provided with a new problem to solve, the first thing students need to do is make sense of what the problem is asking and what information they can find in the math story. Teachers might encourage students to reread the problem, draw a picture, or even underline or write down the key information provided in the problem. For braille readers, making separate notes with the key information may be the most appropriate accommodation. Teachers can also support students in utilizing their reading comprehension strategies to process the context of what a word problem story is describing or asking; they also may encourage students to reread and retell the story problem to a neighbor.
- **Step 2:** Design a plan. As many problems do not specify discrete steps by which to reach the answers, students will need to utilize past problem-solving strategies (e.g., drawing a picture, using tools, acting it out, making a table, etc.) to decide on a plan for how to solve the problem. For students with visual impairments, teachers should encourage them

to use processes that are accessible to them. For example, a student with visual impairments may write out a description of their plan, verbally narrate their steps aloud, use manipulatives, or draw it using the APH TactileDoodle (aph.org/product/tactiledoodle-kit) or Draftsman Tactile Drawing Board (aph.org/product/draftsman-tactile-drawing-board).

- **Step 3:** Enact the plan. Students try to solve the problem using their plan. While they work, they continue to make sure what they are doing is relevant to the word problem context and that the results of their plan make sense.
- **Step 4:** Check their work. Students consider whether their answer(s) are reasonable in the context of the problem. If at any point in the process the work does not make sense, then a student might revisit and retry a previous step (e.g., making sense of the problem or designing a new plan).

Overall, as this practice is a habit of mind, there are no tangible accommodations or modifications that teachers should make for students with visual impairments. Educators should collaborate to ensure that the students have the knowledge and skills to "attack" a problem and are willing to persevere when the solution is difficult to easily determine. While the math educator focuses on students' problem-solving skills and their ability to persist in problem solving, the TSVI has a unique opportunity to support both the practical aspect of solving a problem and the sometimes more emotional aspect of persevering to find the correct solution to a challenging exercise. This emotional aspect of problem solving—which may include believing in oneself, not giving up, and trying a new or untried strategy—connects to the expanded core curriculum (ECC) concept of self-determination.

MP2: Reason Abstractly and Quantitatively

Mathematical Practice 2 (MP2) is about students shifting their thinking from words to numbers and numbers back to words. Students engage in MP2 when they make sense of a word problem and *decontextualize* it—that is, symbolically representing the situation using numbers, variables, and/or equations that correspond to the original story. After finding the answer to their mathematical equations, they then are able to *contextualize* it, plugging that number back into the context of the story. We see elementary students engaging in MP2 when they solve single or multistep word problems using various operations. Secondary students engage in MP2 when they decontextualize a situation by identifying and defining variables and creating an equation to represent a real-world situation using symbols. After building the equation, secondary students can then contextualize the symbolic representation by substituting (or "plugging in") a known variable and understanding how the answer fits within the context of that real-world situation.

Additionally, it is important that students have opportunities to reason in both directions, essentially moving from abstract to quantitative thinking (e.g., decontextualizing a word problem), and also reasoning from quantitative to abstract (e.g., contextualizing a mathematical equation). This means students should have experiences with word problems that they decontextualize, as well as experiences where they must write a math story that accurately matches a given equation.

For students with visual impairments who lack experience with visual contexts, there may be problems that are unclear to them. For example, if a problem asks students to determine the angle of a shadow, students with visual impairments may miss contextual visual information, preventing them from clearly understanding how that concept works in the problem. Therefore, educators will need to ensure that their students with visual impairments receive detailed information so that the problems make sense. Sidebar 1.2 showcases examples of how students might decontextualize a story problem, making sense of the words using manipulatives, drawings, numbers, and symbols. It also shows ways that students might start with an equation, such as $4 + 5 = 9$ or the formula for a linear equation, and contextualize it into their own original story about those numbers or symbols.

Sidebar 1.2 Examples of Contextualizing and Decontextualizing

Examples of Contextualizing

Prompt: Write a story about $20 - 3 = ?$

Student answer:

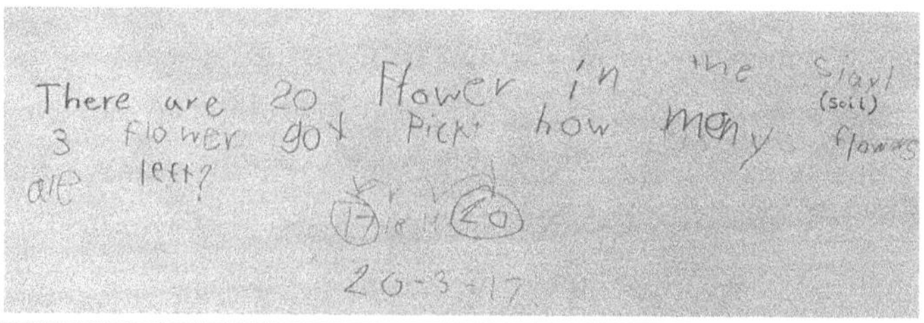

Sidebar 1.2 *(continued)*

Prompt: Consider something in your life that could be represented by a linear equation ($y = mx + b$). Tell the story and write the equation.

Student answer:

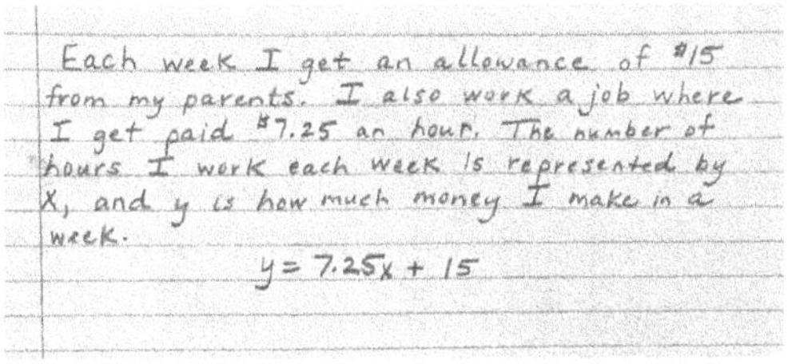

Examples of Decontextualizing

Prompt: There were 4 cars in the parking lot. Five more cars drove in and parked. How many cars are in the parking lot now?

Student answer: 4 + 5 = 9 cars. (For a braille reader, the student could use manipulatives (counters) or make tally marks or full braille cells on the braille device.)

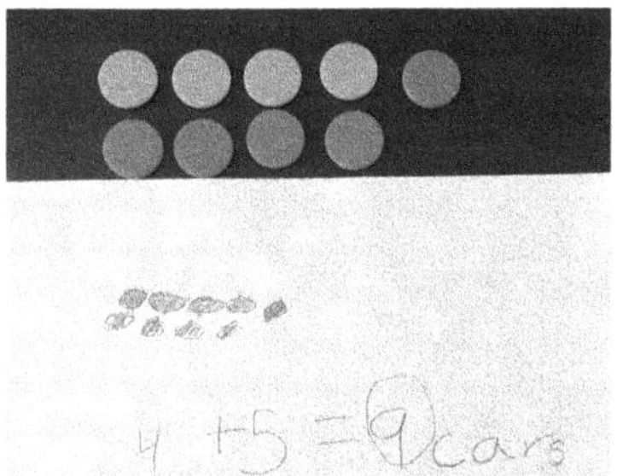

Prompt: Susana rode the bus to get to school. Emil's mom drove him twice as far as Susana rode. Malik rode the bus 3 miles more than Susana did. If the three friends traveled a total of 19 miles, how far did each friend travel to get to school?

Student answer:

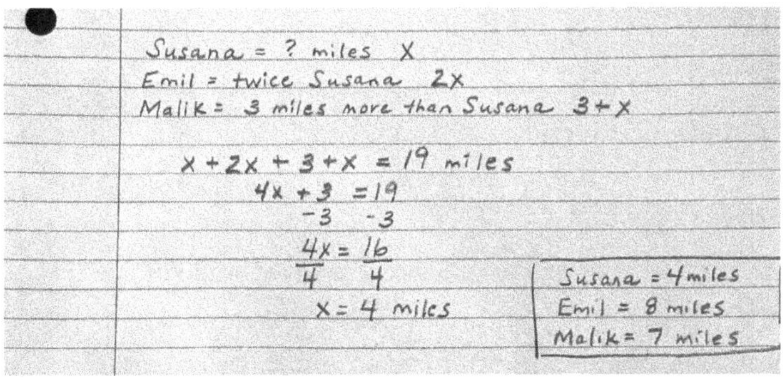

MP3: Construct Viable Arguments and Critique the Reasoning of Others

Mathematical Practice 3 (MP3) focuses on students communicating their mathematical thinking using logical arguments and being able to defend and justify why their thinking makes sense. As with all communication skills, this practice focuses on both speaking and listening practices. Students engaged in MP3 are able to explain to a partner or to the class how they solved a problem and why their strategy worked. They also listen to other students' solutions and ask questions about their peers' thinking to determine whether the answers and strategies their classmates share make sense. Students can use definitions, examples, and counterexamples to strengthen their explanations, but they may also construct an argument through visual and/or concrete representations. This practice lets students know that the expectation always will be for them to explain their thinking and justify their answers. Good mathematicians do not just explain their own thinking but also consider the thinking of others to determine whether the solutions are reasonable.

In math classrooms, we might see students engaged in MP3 around number sense routines such as number talks. During a number talk, students solve computation problems by using mental math strategies that leverage flexible thinking about numbers. (For more on number talks, see Chapter 2, "Foundations of Numbers and Operations.") Students engage by explaining their own solution path, listening to strategies their peers share, and perhaps even comparing the strategies they have used. Designing sections of a lesson where students turn and talk to a partner about their math thinking, work with a group, or share their work with the whole class provides other opportunities for students to commonly engage in MP3.

Students with visual impairments do not typically have an undue impact on a class's ability to regularly incorporate peer communication. There are key benefits of incorporating this practice in the classroom: Engaging in communicating mathematics ideas with peers helps students who are visually impaired strengthen socialization skills, build confidence, and advocate for their ideas. However, like all students, they must have a clear understanding of the terminology and the concepts within the problems in order to communicate effectively, as well as full access to other students' ideas (i.e., teachers should take care that peers are not conveying information only visually). All educators should support this focus on academically appropriate language for mathematics so that everyone in the math classroom uses the same terminology. If appropriate, students with visual impairments may rely on tactile manipulatives to provide an argument. For example, in a problem where students need to add two-digit numbers that will require regrouping ones to make a 10 (or "carrying" one 10 in the standard algorithm), students—including those with visual impairments—might rely on a pictorial representation of the solution, but the students may also use base-10 blocks in lieu of drawing, which provides an alternative, accessible means of communicating the same information.

MP4: Model With Mathematics

Mathematical Practice 4 (MP4) is about students having the opportunity to solve problems in the real world using mathematics—in other words, "modeling" their world. Briars et al. (2013) note that "this practice is often misinterpreted as representing mathematical concepts by using manipulatives" (p. 46). It's important to remember that MP4 focuses on the verb "model"—to create a representation of a real-world situation—and not the noun "model"—a small-scale representation of something. Though students may use physical objects (e.g., using base-10 blocks to build the number 147) when they are modeling, it can be accomplished without manipulatives. Depending on the students' grade level, they may use tools to act out a story problem (essentially modeling the real-world story) or draw pictures. As students get older, they might model data or real-life experiences using tables, graphs, symbols, and equations in order to solve problems. The key to MP4, then, is that students engage in problem solving using appropriate representations to model the mathematical qualities in the world around them.

For students with visual impairments, the action of modeling may require that they utilize manipulatives more than their peers do (particularly for drawing tasks). When modeling advanced mathematics that requires software, accessible software is necessary; for many uses, accessible software is already widely available. For example, if a student wishes to model data using tables, many spreadsheet applications (such as Microsoft Excel) can become accessible once the student receives specific training from the TSVI. Students may complete graphing with a variety of techniques or even create the graph using accessible software and then emboss or print the final product. Later chapters explore specific techniques for modeling and their use in specific mathematical contexts.

MP5: Use Appropriate Tools Strategically

Mathematical Practice 5 (MP5) is about using mathematical tools to solve problems. Students should have access to a variety of grade-level-appropriate mathematical tools and be able to independently decide when and how a tool might be useful to them in solving a problem. This practice is not just about using the tool but also about understanding the affordances and limitations of various tools for various problems. Students should be able to select tools that would help them solve a problem and then use the tools in strategic and efficient ways.

The way a student decides to use a tool can provide insight into their mathematical understanding of the problem and the complexity of their strategies. For example, early elementary students might choose to use connecting cubes or tiles to represent a word problem. While the students engage in MP5 and uses the tools, a teacher might ask questions about the students' models and watch to see whether the students highlight the two parts in the problem using different colors (see Sidebar 1.3). A number line and a hundreds chart are mathematical tools that students might locate in the classroom and use to think about adding or subtracting values. Students might also rely on traditional math tools that also fall under MP5; they may use rulers and protractors for measurement or use a calculator to notice patterns, make an estimate, check their work, or graph equations. No matter what math tool students use, teachers should let *them* decide when and how to use tools independently and in efficient and strategic ways that support mathematical thinking.

For students with visual impairments, teachers can make many tools accessible through employing a variety of colors and textures; in some cases, schools can purchase specialized tools. For example, if students use connecting cubes, adding textured paper or tactile stickers to the blocks should make them accessible. Organizations such as the American Printing House for the Blind (APH) have a variety of tools for measurement, such as specialized rulers with calipers, available for purchase. Further discussion of these tools will appear throughout this textbook.

Sidebar 1.3 Examples of First Graders Using Mathematical Tools to Solve a Math Story

Prompt: There are 9 girls playing on the swings. There are 6 boys by the slide. How many children are on the playground?

Student answers: Student A used connecting cubes and chose colors to represent the parts of the whole:

Student B used tiles without corresponding colors but used positioning to indicate parts of the whole:

Student C used tokens on a grid:

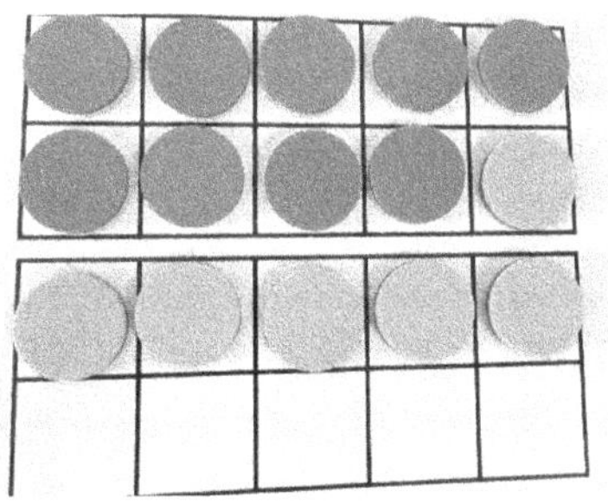

For students with visual impairments, Student C could use tactile tokens, which have felt on one side instead of hard plastic.

MP6: Attend to Precision

Mathematical Practice 6 (MP6) focuses on communicating precisely about mathematics. Briars et al. (2013) describe communication as "includ[ing] oral and written communication, vocabulary, notation, and symbols" (p. 37). Students engaged in MP6 define variables, label axes in graphs, and collaborate to create definitions for mathematical terms so as to communicate clearly. For example, we want students to accurately use the terms *rectangle* and *square* when pointing out shapes in a story, or to use the term *slope* instead of the phrase "the go upped-ness of a line"—which describes slope but is not mathematically precise.

When working with students with visual impairments, the use of precise language takes on additional importance, as some concepts may be unclear or represented differently for students with visual impairments. For example, braille users write fractions horizontally, so if we call the numerator "the top" and the denominator "the bottom," then students who read braille may not be able to apply the spatial concepts presented. The TSVI will need to provide directed instruction on how numbers appear in print and how sighted students communicate about them. At the same time, the TSVI may need to explain to the general educator how braille presents fractions differently. For example, in braille, a fraction includes a fraction indicator that tells the reader that the fraction has "started," the numerator, a fraction bar, the denominator, and then another indicator that ends the fraction, all on a horizontal braille line.

The TSVI will also need to explain to students who use braille that spatial arrangements can impact the procedures needed to solve problems and then prepare the students to adapt the material. As an example, when working with proportions in the form of fractions, the means–extremes property requires students to be able to compute "cross products." Figure 1.1 demonstrates why students who use braille will not inherently understand what is meant by cross products, as they do not write proportions as vertically aligned fractions. Students will need to understand the differences between the braille format and the print format to understand the concepts.

Additionally, all students need to calculate accurately and efficiently, make sure to use appropriate procedures and calculations, and include labels in their problem solving and units of measure (e.g., 4 *apples*, 5 *cm*, 25 *miles/hour*). These labels give meaning to the values and provide clear communication of the solved problem. For students with visual impairments, it is imperative that the TSVI stress the importance of labeling as well as provide access to accessible calculation tools (e.g., talking calculators or an abacus).

Figure 1.1 The Means–Extremes Property Represented as "Cross Products"

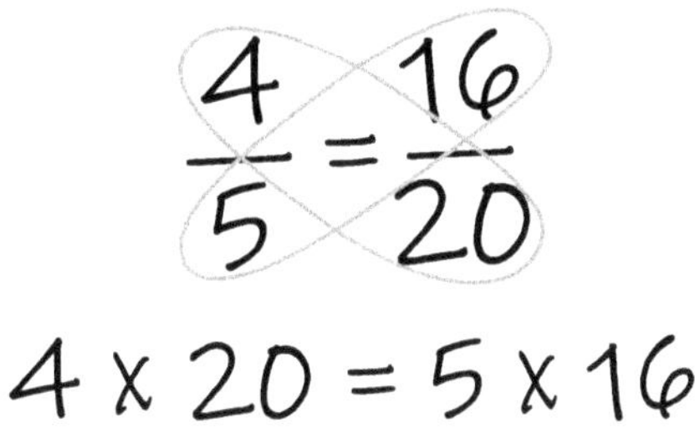

The concept of "cross products" relies on the visual presentation of the proportion and the cross shape that the student produces while working the problem.

MP7: Look for and Make Use of Structure

Mathematical Practice 7 (MP7) is to look for and make use of structure, which encourages students to notice patterns and structures in mathematics. For example, our base-10 number system involves sets of ten that people can identify using place value. Students learn the structure of numbers and find ways to use place value to help them think flexibly. Once students understand that they can break apart a number like 154 into one hundred, five tens, and four ones, they have learned how the structure inherent to place value is a valuable tool when they are mentally adding or subtracting.

Students also notice patterns when exploring operations; for example, understanding $4 + 5 = 5 + 4$. Students recognize patterns when they explore geometry, noticing similarities and differences among definitions (e.g., every square is a rectangle), and by noticing when a particular shape is in a problem, they can use what they know about that shape (e.g., formulas for area or perimeter) to help them solve the problem. There are also patterns in linear and exponential functions, such as rates of change and noticing the direction of a line that has a negative slope. Students will recognize many of these geometric and symbolic patterns (such as the task shown in Sidebar 1.4) as they visually inspect the mathematical outputs.

Sidebar 1.4 Example of a Pattern Task for Engaging Algebra Students in MP7 and MP8

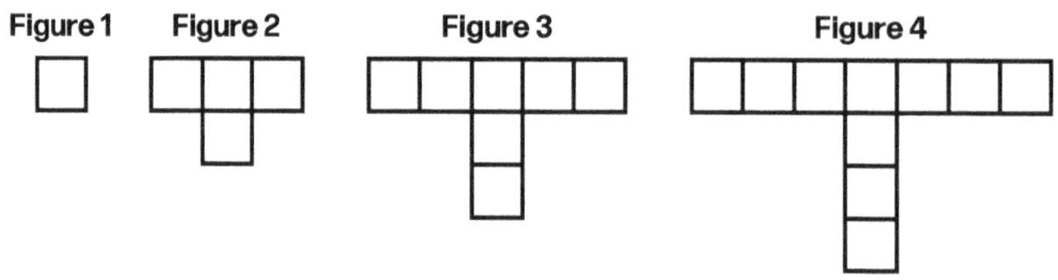

1. Describe how the pattern is growing.
2. Draw the next two figures. How many tiles are in each?
3. How many tiles are in the 10th figure?
4. How many tiles would be in the nth figure? Could you write this as an equation?

This task promotes students' engagement in MP7 during Questions 1–3 as teachers encourage them to notice the structure of the figure as they describe how the structure is growing, and then make use of that structure to draw the next two figures and solve for the number of tiles in a theoretical tenth figure. Question 4 of the task encourages students to engage in MP8 by asking them to think beyond just finding the number of tiles in a particular figure, and instead asking them to develop a rule for any figure. The TSVI will need to create the figures, either as a tactile graphic or by using some other manipulative, such as flat squares.

For students with visual impairments, learning patterns should begin parallel to the ways their sighted peers receive introduction. The APH MathBuilders, Unit 1: Matching, Sorting, and Patterning Kit (aph.org/product/mathbuilders-unit-1-matching-sorting-and-patterning-large-print-kit-includes-teachers-guide-in-print) is an example of an accessible tool that teaches the foundational understanding of geometric patterns to students with visual impairments. This knowledge leads to understanding of numerical patterns within the base-10 number system such as even and odd numbers, multiples, etc. For students with low vision or those who read braille, TSVIs will need to consider where these patterns appear in students' work and might even provide scaffolded questions to support students in noticing these structures when solving problems. Sidebar 1.4 demonstrates an example of this, and either tactile graphics or tactile manipulatives (e.g., tactile squares or APH's StackUps cube kits) can help make it accessible.

MP8: Look for and Express Regularity in Repeated Reasoning

Mathematical Practice 8 (MP8) focuses on students noticing when patterns repeat and using this repeated reasoning to generalize a rule or shortcut. This means students are moving beyond just noticing a pattern (MP7) and are now looking for ways to make their work more efficient. This can lead to the students generalizing the rules, such as discovering the commutative property of addition and understanding that the order does not matter after they have explored various examples. Another rule students might generalize on their own is the formula for the area of a triangle, $A = \frac{1}{2}bh$, which they may discover after finding the area of various triangles on a geoboard and noticing that each has the area of half of a rectangle. The goal of MP8, then, is for students to discover mathematical rules on their own instead of having a teacher tell them.

When working with students with visual impairments, it is critical that the general education math teacher collaborate with the TSVI to find appropriate strategies to support the students' ability to look for and generalize the repeated patterns across problems that lead to discovering shortcuts or rules. This may mean that the math teacher provides the TSVI with more examples and/or problems to allow the students more opportunities to generalize their own procedures, or guided questions that support the students to make connections across the procedures they have been using and develop a rule. Sidebar 1.4 shows a pattern task that contains scaffolded questions to encourage students to move from noticing the structure in a set of figures (MP7) and using this repeated pattern to then develop a rule (i.e., equation) that works for any figure number.

CONCLUSION

The Standards of Mathematical Practice in the Common Core (NGA Center & CCSSO, 2010) outline the ways in which teachers should engage all students in learning mathematics through K–12 education. Students should leverage these eight mathematical practices in every grade, and the mathematical content standards for a lesson will help drive which mathematical practices teachers decide to incorporate. Overall, these practices promote important habits of mind for doing mathematics, and proactively incorporating them into lesson design will encourage students to think like mathematicians.

REFERENCES

Briars, D. J., Asturias, H., Foster, D., & Gale, M. A. (2013). *Common core mathematics in a PLC at work, grades 6–8.* (T. D. Kanold, Ed.). Solution Tree Press.

National Governors Association Center for Best Practices (NGA Center) & Council of Chief State School Officers (CCSSO) (2010). *Common core state standards for mathematics.* https://corestandards.org

National Council of Teachers of Mathematics. (2000). *Principles and standards for school mathematics: An overview.*

National Research Council & Mathematics Learning Study Committee. (2001). *Adding it up: Helping children learn mathematics.* National Academies Press.

Pólya, G. (1945). *How to solve it: A new aspect of mathematical method.* Princeton University Press.

CHAPTER 2

FOUNDATIONS OF NUMBERS AND OPERATIONS

Aina Appova and Tiffany Wild

INTRODUCTION

This chapter focuses on pedagogical suggestions that can help teachers use specific strategies to develop children's foundations of mathematical knowledge in number sense, counting and cardinality, and the base-10 (whole) number system. We particularly place a strong emphasis on the need for meaningful, engaging, and cognitively rich pedagogical approaches and student experiences with mathematics.

Historically, mathematics education for students with disabilities has been taught by emphasizing "telling mathematics" and providing rule-based mathematics learning experiences to children: "Students were taught to solve each problem in a specific single way, and students' informal strategies were disregarded" (Tournaki, Bae, & Kerekes, 2008, p. 41). In fact, a comprehensive literature review of studies on math interventions found that most mathematics interventions focused on teaching procedures (rules, facts, or step-by-step solutions) without much attention to children's ideas, thought processes, and construction of conceptual knowledge (Maccini & Hughes, 1997; Butler, Beckingham, & Lauscher, 2005). As such, this chapter emphasizes approaches that focus on alternative (versus traditional or "standard") methods of performing mathematical operations to help students succeed, make progress in mathematics, and meet grade-level standards and expectations.

Teachers need to focus on teaching mathematics that reflects the needs of the students and incorporates and builds upon their thinking about mathematic concepts. Teachers must understand and appreciate their students' strengths, abilities, and own way of developing conceptual understandings of mathematical thinking to be able to help them develop a strong sense of confidence, accomplishment, and motivation in mathematics. This includes making sense of mathematics problems, developing multiple solutions for mathematical operations, and developing a sense of perseverance in their problem solving. Further, building on a student's own conceptual understandings and abilities to make sense of the world around them gives them the skills to solve problems in a traditional mathematics classroom and in real-world applications.

The teaching techniques in this chapter will help teachers structure their mathematics lessons around students' learning, thinking, and progress at each stage of mathematical development, and they have been adapted and developed specifically for students with visual impairments.

Contrary to the public belief that elementary school involves *easy* mathematics, elementary school actually involves *fundamental* mathematics. The fundamentals are far from easy to teach or to learn. They require a careful and skillful approach to teaching and learning designed specifically to help young learners conceptually understand these fundamental topics and build upon these concepts to develop proficiency in K–12 grades and later in their professional careers.

USING COUNTING PRINCIPLES TO DEVELOP CARDINALITY

Students' experiences with numbers begin very early in their toddler years. In 1930, Lev Vygotsky (1978) described numerous self-evident observations of young children being aware of numbers and able to reason about arithmetic without having any formal instruction on these topics. Decades of research followed his claims, documenting young children understanding numbers long before they attend school, particularly understanding quantities and being able to think quantitatively (Bisanz, Sherman, Rasmussen, & Ho, 2005; Gallistel & Gelman, 1992; Gelman, 1980; Ginsburg & Seo, 1999; Piaget, 1952; Vygotsky, 1978). Some researchers even claimed that children, as early as infancy, appear to be sensitive to and aware of numerical and related spatial representations (Antell & Keating, 1983; Wynn, 1992). Therefore, *cardinality* and *counting principles*—both mathematically and developmentally related—are at the heart of early childhood and prekindergarten learning experiences.

Gelman and Gallistel (1978) identified five counting principles that build the foundation for children to understand base-10 numbers (or whole numbers) and place value (Sidebar 2.1). Structuring mathematics teaching around young students' progressions through these subtopics (and perhaps in this order) can help teachers advance students through each of these progressions more productively, focusing more on the individual student and their thinking, learning needs, and mathematical progressions, all within grade-level standards and expectations.

Sidebar 2.1 Five Basic Counting Principles

The principles outlined by Gelman and Gallistel (1978) demonstrate how young students move through the initial steps of understanding the underlying thought process of counting:

1. **Stable order:** Stable order is the ability of a student to count and say numbers in a sequential order. At this stage, students have memorized the language of numbers in order. For example, they can recite numbers aloud up to 10 from memory without possibly understanding their use.

2. **One-to-one correspondence:** One-to-one correspondence is when students assign just one number to each object of a set. Students will often touch one object, assign it a number, proceed to the next, and so on. At this stage, the students can now assign the number words to objects, one item at a time.

3. **Cardinality:** Cardinality is closely related to counting. Students must understand that the first item is "1" and that the number assigned to the last item represents the total quantity of items in the set. At this stage, the students can now assign a number to a set, which is a fundamental aspect of number sense.

4. **Abstraction (or conservation):** Abstraction is the understanding that the process of counting stays the same no matter the characteristics of the items counted. At this stage, students are able to apply one-to-one correspondence and cardinality to any set of objects,

Sidebar 2.1 *(continued)*

regardless of how the items are organized. This means that they do not assign a lower or higher number based upon differences in spatial size (how close or spread out the objects may be).

5. **Order irrelevance:** Finally, order irrelevance is the understanding that no matter how items are counted (random, left to right, etc.), it does not matter as long as every object is counted only once. At this stage, which may come as late as third or fourth grade, students understand that they can count objects within a group in any way, as long as each item is only counted once. Thus, students clearly understand that they can count in multiple ways.

In early childhood (prekindergarten) years, children learn to verbally count numbers out loud (in a *stable order*). At this stage, children often demonstrate readiness to start counting objects in front of them by assigning one number word to each object and using one-to-one matching (*correspondence*). Children who can correctly count all the objects in front of them and state the total (final) number of objects they counted demonstrate *cardinality*. At this stage, developing *abstraction* and *order irrelevance* is quite easy for children, and teachers should implement additional activities that require children to count different objects, asking children not to focus on colors, sizes, or shapes and counting the objects in a variety of ways (left to right, right to left, randomly).

Sidebar 2.2 Counting Skill Lesson Plans

The National Center on Intensive Intervention at the American Institute of Research (intensiveintervention.org) provides many lesson plans and tips for teachers for counting skills. Lessons often feature videos showing each skill as well as demonstrations of how to assess the skills to ensure students are ready to move to the next level.

Counting Principles for Students With Visual Impairments

Students, including those with visual impairments, are often not explicitly taught the five principles of counting (Phan & Conroy, 2022). Children with visual impairments may have limited conceptual understanding of counting due to this lack of access or exposure. According to Smith and Smothers (2012), this may cause confusion in understanding basic mathematical terminology when comparing objects such as "many" and "few." Students with visual impairments may use different sensory channels to make sense of the world and therefore must be given the ability to tactually and acoustically count different tangible objects (Rottmann, Haberzettl, & Kramer, 2020). However, this may cause students to skip some objects (Phan & Conroy, 2022); it can be challenging to count all of the pieces in a set without seeing the entire set. Also, without a proper understanding of mathematical language, students "may guess the link between the number sequences and the number words without understanding the relationship between the number and the context" (Phan & Conroy, 2022). Young students with visual impairments may also count or repeatedly omit objects entirely unless the objects are organized beforehand; therefore, it is beneficial to create a well-defined area that contains all the objects to be counted. Smaller objects can be placed in a bin or serving tray that has well-defined sides, while larger objects, such as larger balls, can be placed within a circle such as a hula hoop. It is important to give students clear instructions and guidance before they begin counting.

1. **Scanning:** Help the students scan all of the objects by encouraging the students to move their hands across the entire bin, tray, or hoop. This will help them understand where the perimeter lies and where all the objects are located.

2. **Organize:** Help students organize their objects, if they are randomly distributed in the container. This can be as simple as systematically lining up the objects at the bottom or along the side of the container.

3. **Counting:** Students can do this done multiple ways, such as (1) moving the objects from one side of the container to another while counting; (2) picking up individual objects while saying their assigned number and moving them outside the perimeter of the container; or

(3) picking up each item and assigning a number while handing the object to the teacher or while the other hand tracks the next item to be counted (Phan & Conroy, 2022).

Research has shown that in their younger years, children with visual impairments exhibit some differences in development of whole-number knowledge compared to sighted children (Rottmann et al., 2020). Ahlberg and Csocsán (1999) found that young students who are blind would say a number in words but could not quantify what it meant. Students with visual impairments may also have difficulty knowing the right number words to use. If a teacher simply assumes that students understand the names of numbers, opportunities to develop their skills may be overlooked. Therefore, teachers should also have students verbalize each word, and occasionally teachers need to probe students to tell them what each number means (e.g., "Show me eight beads in the tray"). Phan and Conroy (2022) also note that many students with visual impairments do not always understand that the last number in the number sequence represents the total number of objects (cardinality). Therefore, probing by asking students how many total objects are in a set is vitally important to ensure understanding of counting. It is important for teachers to directly teach one-to-one correspondence to ensure that students understand cardinality; modeling using proper mathematical words in the classroom and giving explicit instructions helps students understand the foundations of mathematical literacy.

Sidebar 2.3 Incorporating Counting Activities Into a Child's Daily Life

Counting should be incorporated into the daily lives and routines of young children as they play. For students with visual impairments, this means having access to tactile items to count with a one-to-one correspondence. Some examples:

- At the grocery store, show the child the location of the item that is needed. Have the student count the items when placing them into the basket.
- When putting away groceries at home, have the student announce the number of items being put away in the pantry or fridge.

Sidebar 2.3 *(continued)*

    - For example, "We bought three boxes of macaroni and cheese today. Can you please hand me the three boxes? Be sure to count them to make sure we have enough."
    - Encourage the child to touch and count each box.

- When the child removes toys from the toy box, ask, "How many toys have you taken out of the box? Can you please count them as you put them away?"
- Ask students to be part of the daily lunch count or attendance. By using a tactile chart, students can count the number of students in attendance.
- At snack time, count the number of items placed on the plate for the child. Again, encourage the child to verbalize as they count.
- When setting the table for meals, ask the child to count the number of napkins, spoons, plates, etc. that they are placing on the table and how that relates to the total number given by the teacher or adult guardian.

For additional examples of counting at snack time or in routines in the classroom, videos from Project Math Access (Early Childhood 2A and 2B) are available at youtube.com/@projectmathaccess2562/playlists.

Asking probing questions when students are beginning to count is very important to ensure all students—especially those with visual impairments—understand the conceptual understanding of cardinality. When giving students a bowl and pom-poms, asking a question such as "How many pom-poms do you think we can get into this bowl?" allows teachers to probe for understanding of counting; it also provides opportunities for problem solving and is a precursor to understanding scale (a concept that can be difficult for students with visual impairments). Other probing prompts could include "Show me how to count, please" or "Are there more chairs than we need for all of our friends? How do you know?" These questions all allow a teacher the opportunity to quickly assess students' understanding and reinforce mathematical literacy.

Teachers should also include experiences that require students to organize objects into specific groups, perhaps at first by nonmathematical attributes (for example, texture or theme) and gradually require students to organize the objects by size and shape and sort a specific number of objects into each group. This allows students to gain skills in mathematics that are required for other subject areas such as science, and it is part of visual efficiency in the expanded core curriculum for students with visual impairments. For example, the teacher can put pom-poms in a jar and ask students to count them and then group them by size or color, if appropriate for the students in the class. This allows students to practice counting and grouping in the same activity. Another great resource for engaging children with visual impairments in early childhood numbers and counting is Lots of Dots (aph.org/product/lots-of-dots-coloring-the-garden-1-10002-00), which specifically covers whole numbers/integers and includes plastic stencils with large braille cells and foam dots to aid in the instant recall of numbers.

Once students, including those with visual impairments, have learned the principle that one number equates to one object, it is important to move on to the next stage. By counting and organizing objects into specific groups, students develop an understanding that their groups make up the total number of objects; this prepares them for *subitizing* and *benchmarking*.

USING CARDINALITY TO DEVELOP SUBITIZING AND BENCHMARKING

Subitizing

Conducting exercises in counting and cardinality naturally prompts young students to create groups of objects. *Subitizing* (the ability to accurately identify the total number of items using groupings instead of counting) is the next step in the process, as it involves grouping by numbers instead of attributes. Teachers engage students in structured opportunities for students to subitize by asking them to represent a certain number of objects in a variety of ways, using different groups. For example, a teacher could give a student six blocks. The teacher would then ask the students to sort the blocks into two equal groups. At first, a teacher would want to model this with students and ask the class for ways to make two equal groups. Once students understood that they should sort blocks into two equal groups, then the challenge would become to make three equal groups or make any variation of groupings that all add up to six.

The next challenge may be to come up with ways to place nine blocks into two or more groups of equal and/or unequal groups. Students at this stage need to show understanding that rearranging the objects into different groups does not change or otherwise affect the total number of objects (Gelman & Tucker, 1975; Gelman & Gallistel, 1978). As the teacher works with the students to develop an understanding of grouping, they must also include the concept of zero by demonstrating an "empty" group. To assist in this, teachers should focus math activities on the use of manipulatives and objects that students can arrange and rearrange into different groups.

As with counting and grouping exercises, a container or tray is a great option for teachers to ensure that students with visual impairments can easily find all the objects and then contain the objects that they will be moving around. A tray with multiple inner compartments, such as a cafeteria tray, works particularly well: The total number of objects can go in the larger compartment, and when the student wants to group objects, the smaller compartments can hold the objects and serve as a tactile marker for each group. Again, the concept of zero can be explored by having the students create groups that do not include any objects that represent zero. APH also has various products for sorting, such as their Work–Play Trays (aph.org/search-results/?fwp_search_term=work+play+trays) that students can use for sorting activities. The APH MathBuilders, Unit 1: Matching, Sorting, and Patterning Kit (aph.org/product/mathbuilders-unit-1-matching-sorting-and-patterning-large-print-kit-includes-teachers-guide-in-print) is a supplementary math program for K–3 that focuses on using tactile objects to build prenumeracy skills.

The skill of writing numbers symbolically in Unified English Braille (UEB) Math/Science or Nemeth code takes several years for students to develop (see Sidebar 2.4). Specific instruction on ways to create the symbols on the braillewriter are required and follow much of the same methods as when introducing literary code. Therefore, teachers must provide students with other opportunities and methods for representing quantities and communicating their numerical reasoning. These activities can easily be adjusted to include students with visual impairments by using tangible items such as plastic counting animals, tiles, or other small tangible objects. Several commercial products are also available on the market for teachers to use with children with visual impairments (see APH's Tactile Treasures Kit: Math and Language Concepts for

Young Children with Visual Impairments, Tactile–Color Edition, available at aph.org/product/tactile-treasures-kit-tactile-color-edition).

Sidebar 2.4 Introducing Braille Mathematics

For students who will need to use Nemeth code or UEB Math/Science, early exposure to a braillewriter and mathematical symbols in kindergarten (or, if possible, prekindergarten) is crucial, and practicing counting is a good opportunity to introduce students to the idea that there are braille symbols that represent numbers and mathematical expressions. They should also begin to learn how to create numbers and mathematical symbols (such as the plus, minus, and equal signs) in Nemeth code or UEB Math/Science at the same time as their peers learn to write them.

To begin, it is important to have students understand the symbol (number word) and the quantity (number of items). One way to do this is with written cards with a mathematic number written on it. Tell students the meaning of the symbol and match the objects with the symbols. For example, providing the students with the symbol written in UEB Math/Science or Nemeth code with the number 2 and then two objects placed beside it. Another way to do this is to take a Styrofoam cafeteria tray or carryout container and cut it so only two sections are included. In the top section, place the card; in the bottom section, place the number of objects. Encourage the students to continue to explore this until they begin to understand the symbols. Research has shown that for students with visual impairments, this method has shown promising results over time (Ivy & Hooper, 2015). High praise and motivation are the key to making this successful.

Benchmarking

When young students begin exhibiting noticeable progress in grouping and making explanations of their groups, they are showing readiness to (formally) be introduced to *benchmarking*: grouping objects into groups by 5s and 10s. The learning goal for benchmarking is to represent a number in very specific ways that use groups of benchmark numbers, as opposed to subitizing, which emphasizes the variety of ways a student can represent a number. Because of this, it is important to use the correct manipulatives for each learning goal: When students are learning to group objects into benchmark numbers (5s, 10s, 100s, etc.), they should be introduced to manipulatives that allow them to *combine* quantities, e.g., an abacus or linking cubes. Once students demonstrate mastery in grouping objects, the teacher can introduce new objects used for grouping that focus on *separating* and *organizing* quantities into benchmark groupings, such as five frames and ten frames. This will help students develop important *mental math* strategies as they develop their benchmarking skills.

An abacus (Figure 2.1) is a wonderful tool that can help students with visual impairments learn counting, and introducing it early prepares them to compute mathematical equations with this tool when they become older. The abacus has been used by various cultures as a calculating tool for centuries. However, it is not a "calculator" in that it does not complete operations for the user. Instead, it is more akin to "scratch paper" in that it allows the user to "set" numbers and then manipulate the beads as the user completes computations. While there are many different types of abaci, the most common abacus used in North America is the Cranmer abacus (aph.org/product/cranmer-abacus).

Figure 2.1 Cranmer Abacus

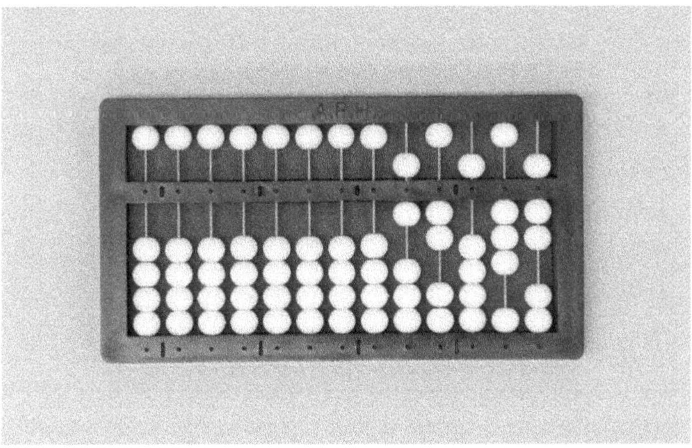

An abacus is a versatile tool that can help students with visual impairments learn to count and, later, calculate.

The Cranmer abacus uses a series of beads as "placeholders" for numbers. Across the top are singular beads. Each of those beads has the value of 5. The second row has a series of five beads, each with the value of 1. The top and bottom rows meet in the center to represent numbers. For instance, sliding one bead to the center represents the number 1. Sliding the top bead down to the center and one bead up to the center becomes the number 6. Moving three beads up from the bottom to meet one bead from the top creates the number 8. As the Cranmer abacus benchmarks 5s and 10s at its foundation, its use can help students develop these fundamental numeracy skills as they progress in mathematics.

The APH Hive (aphhive.org) has modules that provide instruction on using the abacus for all levels of mathematics. For this chapter, the course titled "The Abacus: A Million Manipulatives in Your Pocket," found at aphhive.org/#/courses/enroll/2265 (login required), is designed to help teachers understand the abacus and teach the use of the abacus to their students. Other resources for practice are the Practice2Master Abacus app (aph.org/product/practice2master-abacus) and Abacus Basic Competency: A Counting Method (aph.org/product/abacus-basic-competency-a-counting-method-large-print), which are fully accessible and can supplement instruction to help students practice and strengthen their counting, combining, and addition and subtraction skills.

Initial benchmarking by groups of 5s and 10s is critical for building a student's mathematical (foundational) knowledge required for understanding place value, base-10, and operations on numbers. This understanding expands into upper elementary grades and develops students' learning from being able to represent the numerical values of objects into an ability to analyze and reason about properties of numbers (Gelman, 1980).

The use of five frames and ten frames (Figure 2.2) helps students develop a "frame of mind." They begin to understand that some numbers can only be represented in combinations of 5s, 10s, and plus/minus a leftover amount. They also begin to reason about "double" and "triple" groups of 5s and 10s. All of these strategies are required to provide students with opportunities to develop mental math strategies and proficiency in applying operations (addition, subtraction, and early multiplication) on numbers up to 30.

However, when teaching benchmarking, teachers must recognize that using five frames and ten frames beyond three groups of 10 (numbers greater than 30) is not easily manageable for young children. By the end of first grade, students are expected to be proficient in both subitizing and benchmarking numbers up to 30. This is because the "teen" numbers, especially those that require combining two or more addends that are greater than 5 (for example, 7 + 8), require students to develop addition or subtraction strategies, which moves beyond counting.

Figure 2.2 Five and Ten Frames

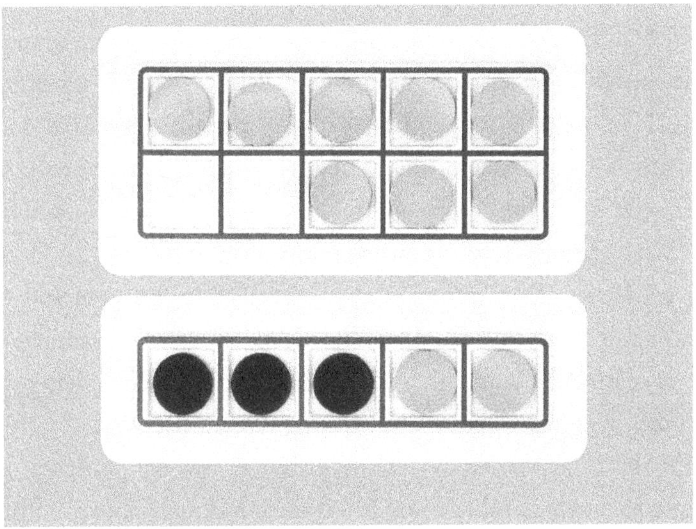

Five frames and ten frames can help students tactually experience quantities and their relationships.

A helpful tool can be APH's Accessible Five and Ten Frames, which can be found at aph.org/product/tactile-five-and-ten-frames. This kit provides teachers with accessible ten frames, five frames, and tactile tokens to help with counting and cardinality skills, place value, addition and subtraction, and methods for composing and decomposing of numbers. The APH Hive offers an online course titled "Tactile Aids for Mathematics" that helps describe how to use these in the classroom and is available for free at aphhive.org/#/courses/enroll/7 (login required). Additionally, ten frames can also be made accessible for students with visual impairments using tactile boxes created with either Wikki Stix (wikkistix.com)or with tactile lines using pipe cleaners, capsulated paper, or other tools. Plastic counters or tokens can then be placed into each tactile frame to allow students with visual impairments to access the images. As always, tactile representations are also beneficial and fun to use with sighted peers in the classroom and thus allow for full participation by all students. Additional videos for using the abacus in the classroom can be found at Project Math Access, available at youtube.com/@projectmathaccess2562/videos.

Addition and Subtraction Strategies

Teachers should teach both addition and subtraction strategies when working with five and ten frames. The "plus" strategies involve students solving 7 + 8 by making a full ten frame plus the leftover amount (for example, 7 + 8 = 7 + 3 + 5 = 10 + 5). It is important to emphasize to students that subitizing either 7 or 8 in this case would be both correct and appropriate to make a full ten frame, and that the leftover amount (5) will always be the same no matter what (e.g., 7 + 8 = 7 + 3 + 5 = 10 + 5 or 7 + 8 = 8 + 2 + 5 = 10 + 5). In contrast, the subtraction strategies would involve students, when solving 7 + 8, to compare 7 and 8 to a full ten frame and recognize that 7 is "3 away from 10" and 8 is "2 away from 10," so the answer must be "5 away from 20." It is imperative for teachers to keep asking students to develop multiple and different strategies to solve the same problem. The ultimate goal is for students to have a variety of mathematically different thinking and problem-solving strategies for any addition or subtraction problem they encounter.

By Grade 2, mathematics activities should involve students counting and writing the total number using five frames and ten frames and providing expressions/equations using plus or minus signs to represent how they got the answer. When they are subitizing, students often represent 7 as 4 on the top frame and 3 on the bottom, reflecting on their knowledge of both subitizing and possibly doubles (e.g., 3 + 3 + 1 or 4 + 4 − 1). When benchmarking, students will represent the same number (7) as a full five frame plus 2, and the written equation will contain "5" in the answer (e.g., 7 = 5 + 2). As students begin exhibiting progress in both subitizing and benchmarking (for the same math problems), teachers should move students one step further by providing extensions. For example, students should be asked to represent the same number (7) using the number 10 in their explanations and equations. A child might discuss 7 as "needing 3 more to make 10" or being "3 away from 10" and write 10 − 3 = 7. These relationships are also known as "fact families.". These questions and activities help to deepen children's mathematical development of number sense and understand numbers and operations beyond counting them.

Sidebar 2.5 Writing and Understanding the Addition and Subtraction Symbols for Students with Visual Impairments

As students practice addition and subtraction, they become more comfortable using the plus, minus, and equal signs. For students who need it, it is important to practice knowledge of those symbols in either Nemeth code or UEB Math/Science on the braillewriter. As before, a teacher may want to use trays or small dishes for this task. Three small, shallow dishes should be placed in a line with three small shallow dishes below, creating two parallel rows. The teacher will place tactile objects representing the first number in the first dish, leave the second dish empty, and then place tactile objects representing the second number in the third dish. A card with the corresponding number written in Nemeth code or UEB Math/Science code should go below each corresponding dish, with the symbol used for the mathematic equation placed in the middle.

For example, if the equation is 10 + 3, the teacher will place 10 tactile objects in the first dish in the first row, skip the middle dish, and place three objects in the last dish on that same row. On the bottom row, a card with the number 10 (written in code) is placed below the first dish, a card with a plus sign (also written in code) in the middle dish, and a card with the number 3 (again, written in code) in the last dish. Once the student begins to understand the braille symbols and is asked to solve the equation, two additional dishes should be added to the top and bottom. At the top, these dishes remain empty. At the bottom, the first additional dish has the equal symbol written in Nemeth code or UEB Math/Science; the last dish is left blank for the student to put the corresponding tactile objects in the top row and the corresponding number in the second row.

Additionally, LEGO Braille Bricks (aph.org/lego-braille-bricks) can also be beneficial to helping with teaching writing and reading equations. Each LEGO Braille Brick has the number represented on the brick in both print and UEB Math/Science or Nemeth code. The student can find the numbers they need and place the bricks in an equation pattern on the table or onto a baseplate. These bricks can be a fun tool for sighted students in the classroom as well. The APH Hive has a module on using LEGO Braille Bricks in mathematics, "Brick-by-Brick Numeracy," which is available at aphhive.org/#/courses/enroll/2131 (login required).

Fact Families

Once students show great progress with subitizing and benchmarking, their experiences should be expanded to extended fact family discussions (up to 20) to help them develop and demonstrate proficiency in fact-family strategies, including solving addition and subtraction problems. A *fact family* is a group of three numbers that can be associated with one another through operations such as addition/subtraction or multiplication/division (e.g., 6, 7, and 13 can be represented as $6 + 7 = 13$, $7 + 6 = 13$, $13 - 7 = 6$, and $13 - 6 = 7$). At this stage of learning, once students display noteworthy progress with addition/subtraction within 20, teachers can start adding "mental math" strategies, which include recall of basic facts by using flash cards. For students with visual impairments, flash cards with large print or the Nemeth code or UEB Math/Science symbols are available, such as APH's Math Drill Cards available at aph.org/product/math-drill-cards-number-and-math-sign-cards-nemeth (Nemeth) or aph.org/product/math-drill-cards-number-math-signs-ueb (UEB Math/Science).

Throughout Grades 1–2 mathematics, students expand on their knowledge through different equation developments in addition and subtraction, including story problems associated with those equations, providing more problem-solving and thinking strategies that help students develop solutions. These experiences are vitally important to the continued development of mathematical literacy.

Students with visual impairments may focus on declaring an answer without using any procedure or calculation in observable ways (Ahlberg & Csocsán, 1999). Therefore, it is vital that students are asked to explain their reasoning and how they solved the problems. This allows teachers to quickly and informally assess student knowledge of basic facts, problem-solving, and thinking strategies—their true mathematical literacy.

Addition and Subtraction Operations

Research has indicated that students with visual impairments group numbers in 2s, 3s, 4s, or 10s, or by using derived facts when performing observable operations (Ahlberg & Csocsán, 1999). This finding is similar to other research that studies children's mathematics skills within four different structures related to addition and subtraction: *join*, *separate*, *compare*, and *part–part–whole* (Carpenter, Fennema, Levi, & Empson, 1999).

Sidebar 2.6 Four Structures Related to Addition and Subtraction

- **Join:** Addition, "joining" numbers (or sets) together
- **Compare:** Comparing quantities (or sets) to find the difference between them
- **Separate:** Subtraction, "disjoining" numbers, removing a set from a larger set
- **Part–part–whole:** Representing the relationship between a whole and its parts. This representation can be made up of two or more smaller parts; combining those parts makes the whole (e.g., 5 + 5 + 3 = 13).

Each of these structures involves a number fact family (e.g., the relationship between 7, 6, and 13) and produces different results depending on which of the three numbers in the structure is unknown. For example, *join* problems always involve the initial amount, the change amount (amount being added/joined), and the result amount. *Result unknown* story problems (e.g., 6 + 7 = ?) are prevalent throughout most mathematics instruction and include story problems such as, "Jayden had 7 toys. Maya gave him 2 more. How many toys does Jayden have now?" Many teachers may attempt to help struggling students by removing any extra wording in a story problem (which often removes the "thinking" and reading comprehension opportunities as well) and will focus on teaching students specific "keywords" to listen or look for when problem solving, including asking students to underline those words when they read the story problem. The keywords taught for addition often include "more," "added to," "total," "in all," and "how many."

Unfortunately, this focus can be a detriment to students, as it takes away critical opportunities for students to learn to read, listen, and make sense of mathematics problems, while preventing them from considering the context and developing problem-solving strategies. It can also teach students strategies (by underlining the keywords) that sometimes lead them to incorrect solution paths (see the *Cognitively Guided Instruction* (CGI) project; also see Carpenter, Fennema, & Franke, 1996; Fuson & Fuson, 1992; National Research Council et al., 2001).

Consider the following story problem (from Van de Walle, Karp, Lovin, & Bay-Williams, 2013, p. 65) and the strategy of solving it by underlining the keywords:

> Aiden had 7 marbles in her bag. After school, she found some more marbles that she had misplaced and added them to her bag as well. When she counted all of her marbles, there were 15 in all. How many marbles did Aidan find and add into her bag?

Most students who were taught to underline keywords would be inclined to solve this problem (incorrectly) by adding 15 + 7. This particular problem has the *join change unknown* structure, meaning it defines/states what the initial addend and the sum are and asks students to find the unknown number that was added/joined. By focusing on keywords, students will miss what is being asked of them.

One more structure in the join category, called *join initial unknown*, includes story problems that define/state the second addend and the sum but keep the initial addend unknown. For example, "Sandra had some pennies. George gave her 4 more. Now Sandra has 12 pennies. How many pennies did Sandra have to begin with?" (Van de Walle, Karp, Lovin, & Bay-Williams, 2013, p. 68). This structure seems to be most difficult for children, perhaps because most of their experience includes being told what the first addend is to help them "add to" it or use it as a reference when solving the problem. These problems directly align with the Operations and Algebraic Thinking Common Core State Standard (CCSS.Math.Content.1.OA.D.8) that suggests including these structures in children's mathematics learning. The learning goal (CCSS, 2010, p. 15) states that students should be able to

> [d]etermine the unknown whole number in an addition or subtraction equation repeating three whole numbers. For example, determine the unknown number that makes the equation true in each of the equations 8 + ? = 11, 5 = ? – 3, 6 + 6 = ?

Similar problem structures are taught for subtraction, as *separate* problems, including separate result unknown (most taught in the United States), *separate change unknown* (sometimes taught in the United States), and *separate initial unknown*, a more difficult problem type that is taught less frequently in the United States. Two additional structures, representing addition and subtraction, include the *compare* problems and *part–part–whole* problems, where either the "whole" is unknown (e.g., "Jayden has 6 red marbles and 5 green marbles. If he puts all the marbles in a bag, how many marbles will he have total?") or the "part" is unknown (e.g., "Jayden has 11 marbles. Of these marbles, 5 of them are green and the rest are red. How many red marbles does Jayden have?").

The *compare* problems are easier for children in comparison to other addition and subtraction problems, as they always have two numbers to compare and students often do this (successfully) by using linking cubes (Figure 2.3) to represent each quantity and compare their size and length. The *difference unknown compare* problems often include the following structure: "George has 12 pennies and Sandra has 8 pennies. How many more pennies does George have than Sandra?" (Van de Walle, Karp, Lovin, & Bay-Williams, 2013, p. 69). Other compare problems typically include mathematical structures that include wording such as "more than" or "greater than" or "fewer than" or "less than"—for example, "George has 4 more/4 fewer pennies than Sandra. Sandra has 8 pennies. How many pennies does George have?" Other *compare* examples may include, "George has 4 more/4 fewer pennies than Sandra. George has 12 pennies. How many pennies does Sandra have?" (Van de Walle, Karp, Lovin, & Bay-Williams, 2013, p. 69).

Figure 2.3 Linking Cubes

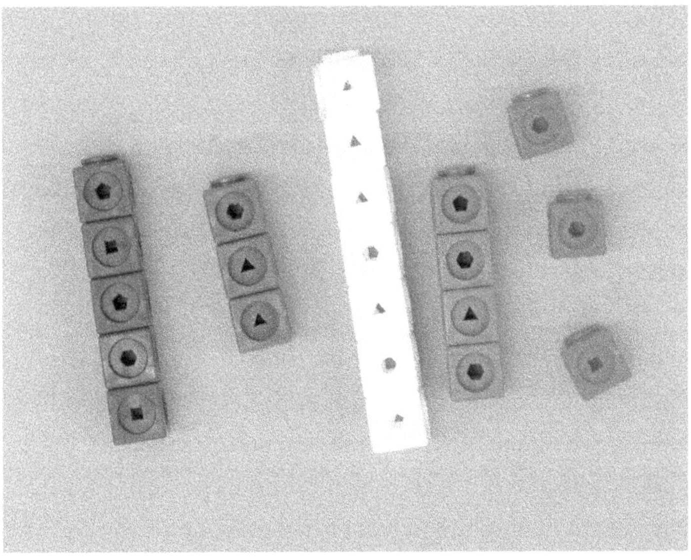

Linking cubes allow students to physically join cubes together to represent values.

Mathematical Thinking Strategies

The CGI project (Carpenter & Fennema, 1992; Carpenter, Fennema, & Franke, 1996; Carpenter, Fennema, & Franke, 1999; Moscardini, 2014) documented that children develop number sense and understanding of operations on whole numbers via three levels of thinking strategies:

- **Level 1:** Direct modeling, concrete manipulatives, and one-by-one counting
- **Level 2:** Skip counting, addition/subtraction strategies (e.g., skip counting by 2s, "counting on" [starting on a number and counting from there], benchmarking ten(s) then adding/subtracting the leftover amount, etc.)
- **Level 3:** Mental math strategies, additive reasoning, derived facts (doubles, doubles plus/minus 1 or 2, doubles "near" tens ($9 + 9 = (10 + 10) - 2$), doubles "near" fives ($6 + 6 = (5 + 5) + 2$), etc.)

At each level, children continue to make (mathematical) progress toward mastery using specific strategies until they reach Level 3, a full proficiency level using mental math (Carpenter, Fennema, & Franke, 1996; Fuson & Fuson, 1992).

At Level 1, children often use counting objects or their fingers for the numbers that they are adding and subtracting, directly modeling their thinking.

At Level 2, children use counting strategies that continue to utilize counting but are often coupled with a "developing" mental math strategy demonstrated by their knowledge of mathematical strategies. Also at Level 2, children no longer need to represent and count both quantities; instead, they use strategies such as "starting at 7 and counting on" for addition. At Level 2, they may recognize that $7 + 5$ is about 7 being "3 away from 10" and use *10-frame benchmarking* with a "counting on" strategy as a solution path: $7 + 5 = 7 + 3 + 2 = 10 + 2 = 12$. For the same problem, at Level 2, children may recognize that "double $5 = 10$" by using *5-frame benchmarking* and use a "counting on" strategy to arrive at a different solution path: $7 + 5 = 5 + 5 = 10 + 2 = 12$. At this level, children actually do not memorize addition facts such as $7 + 5$, so using flash cards to teach addition and subtraction is not helpful. Instead, children use a variety of subitizing and benchmarking strategies to help them decompose the numbers into "easier" addition and subtraction problems, creating more "familiar" number structures and fact families.

Level 3 is associated with mental math. Research suggests that teachers often move too quickly from Level 1 to Level 3 and teach children fact recall and memorization, erroneously assuming that mental math (Level 3) is entirely based on children memorizing the facts. Doing so bypasses students' learning opportunities to develop counting strategies, number images, and quantitative reasoning (Level 2), which are essential and required perform mental math computations (Level 3) (National Research Council et al., 2001). Therefore, we encourage teachers to provide students with ample opportunities to continue developing multiple strategies and demonstrate progress in gradually becoming more comfortable with decomposing numbers as part of these strategies; along the way, students will be increasingly ready to do the "decompositions" mentally, without the need to "count" objects or fingers any longer. Teachers can advance students further by providing opportunities for them to recognize "doubles" and "count on" and "plus/minus 1 or 2" as part of mental math strategies and discussions.

Using Number Talks in the Classroom

One of the most effective pedagogical methods to implement mental math discussions in math classrooms is called *number talks*. A number talk is a mathematical conversation (10–15 minutes long) focused on purposely crafted, teacher-posed computation problems as a tool to structure the classroom instruction around the essential process and habits of mind associated with doing math mentally. Says Parrish (2011, p. 199), "During number talks, students are asked to communicate their thinking when presenting and justifying solutions to problems they solve mentally. These exchanges lead to the development of more accurate, efficient, and flexible strategies." A successful number talk is rooted in basic children's talk, with the teacher asking them to communicate and share their computation strategies and letting students do most of the talking. The teacher writes a computation problem on the board (e.g., 26 + 37), gives students time to mentally solve it, and then each student quietly puts a thumb to their chest to let the teacher know they have an answer and are ready to share their strategy. After several students share answers and strategies, the teacher encourages the students to continue thinking about other ways and more efficient strategies.

A key to a successful number talk is building a type of classroom environment that is inviting, safe, and risk free. It is focused on making the students comfortable in sharing responses, eager to partake in all discussions, and unafraid to question themselves and their peers. This type of classroom environment takes time and effort, as teachers must work with children on learning how to become more accepting of different ideas (and answers) based on a mutual quest for helping one another learn and understand mathematics. Teachers can set this expectation from the beginning; again, Parrish (2011, p. 202) maintains, "a first step toward establishing a respectful classroom learning community is acceptance of all ideas and answers—regardless of any obvious errors. Rich mathematical discussions cannot occur if this expectation is not in place." The beauty of number talks is in the opportunities it offers children to continue engaging in mathematical conversations across numerous topics, including large numbers, negative numbers, decimals, fractions, irrational numbers, and operations on numbers.

USING BENCHMARKING TO DEVELOP BASE-10 NUMBER SYSTEM AND OPERATIONS (INTRODUCING ZERO AS A PLACE VALUE/HOLDER)

Addition and subtraction mental math strategies for numbers up to 100, as well as benchmarking by 10, are foundational knowledge necessary for children to develop understanding of the base-10 number system, its structure, and operations on multidigit whole numbers. Early experiences of children making sense of base-10 numbers include benchmarking by 10 and "regrouping" into groups of 10 and 100 for multidigit (three-digit) numbers. These experiences formally introduce students to the concept of place value and a unit (one) to define multidigit numbers as numbers that have a base-10 value in each place with only 10 possible digits: 0, 1, 2, 3, 4, 5, 6, 7, 8, and 9. A widely used mathematics manipulative (tool) that represents this concept are base-10 blocks (Figure 2.4). Base-10 blocks come in different colors, including different colors for each place value, and have specific "labels" for each place value: unit, rod, flat, and cube. Many vendors, such as APH, offer base-10 blocks as part of their mathematics manipulatives kits (see aph.org/product/focus-in-mathematics-base-ten-blocks-flats-10-per-set).

Teachers must keep in mind several aspects that seem to impact children's learning with base-10 blocks. First, teachers should provide base-10 blocks to each student, otherwise the tool is not effective. Also, when providing students with the tool, it may help to not tell them right away what the rod or flat represents, as it is more effective to have students explore the tool and first describe each piece in terms of shape, size, and unit-based quantity. It is also helpful to begin every base-10 block experience with a mathematical conversation, asking students to describe (and remind everyone) what different pieces mean, give their respective names, and define how many units are in a rod and a flat. One word of caution: Teachers often bring out the cube and tell students that a cube equals 1,000 units, which can lead to confusion, as students typically are not taught the concept of volume until Grade 5. Concluding that a cube = 1,000 units requires students to know that there are "invisible" units inside the cube's dimensions, whereas students may only apprehend the visible units on the faces of the cube, creating a misconception that a cube represents 600 units (100 for each face). To counter this, teachers have two options: They can take the opportunity to discuss/teach volume of a cube informally, or they can choose to develop children's understanding of base-10 numbers in earlier grades only using units, rods, and flats—not cubes. We advise teachers who would like to informally introduce/teach the volume of a cube to specifically use interlocking base-10 manipulatives (see the blue base-10 blocks in Figure 2.4) by asking students to stack/interlock the bases (flats) as high as they need to reach 1,000 (height = 10). This informally introduces students to the volume formula (area of base × height) and helps them tactually build each base (flat) and count each 100 until they reach 1,000.

Figure 2.4 Base-10 Blocks

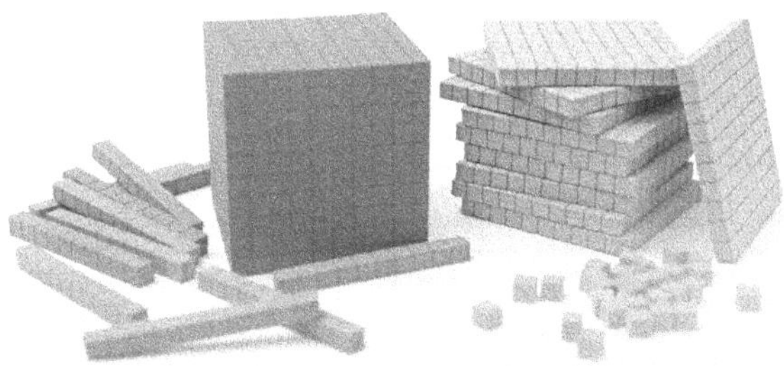

Manipulatives such as base-10 blocks use color and shape to differentiate place values.

Introducing the Concept of Zero as a Place Value/Holder

At this stage, it is critical to formally introduce children to *zero as a place value/holder* by reasoning about three-digit numbers that contain a zero, such as 430 or 403, and asking students to explain what zero represents in each number and how/why these two numbers may look similar when written out in numerals yet are different in quantity and when represented with base-10 blocks. Students should be able to explain which number is larger/smaller using words such as *hundreds* and *tens place*. By Grade 2, students are expected to understand place value and be able to write multidigit numbers in *expanded form* (e.g., 476 = 400 + 70 + 6). Writing numbers in expanded form helps students to learn to add/subtract multidigit numbers using place values (e.g., 349 + 285 can be represented as 300 + 200, 40 + 80, and 9 + 5) and fully demonstrate the understanding of zero as a place value/holder. Teachers should include opportunities for students to write multidigit numbers in expanded form and should begin to introduce parentheses to students' vocabulary and mathematical work, which helps keep the operation for each place value in the expanded form separate: 349 + 285 = (300 + 200) + (40 + 80) + (9 + 5). Rendering mathematical problems in this way prepares them to use *partial sums* or *partial differences* to complete their work, which is a valuable tool discussed in the next section.

To provide additional support for developing arithmetic skills, teachers can engage children in exploring the "hundreds chart," which is a 10-column, 10-row table containing the numbers 1–100. The hundreds chart helps students notice and explore number patterns, count by 10s and 1s, and develop mental math strategies for addition and subtraction within 100. A great resource to use for these explorations is the Hundreds Board and Manipulatives (available in Nemeth code and UEB Math/Science, respectively, at aph.org/product/hundreds-board-manipulatives-nemeth and aph.org/product/hundreds-board-manipulatives-ueb), which includes counters, numbers, and shapes that can be used with the 1–100 grid board (including the Consumable Hundreds Chart, aph.org/product/consumable-hundreds-chart) to help learners become flexible and fluent problem solvers for numbers within 100. Another option is the Place Value Setter (aph.org/product/place-value-setter), which is specifically designed for Grade 1–3 students to use as they learn to represent large numbers using place value and written digits. By Grade 3, students are expected to no longer have the need to use base-10 blocks to write multidigit numbers or perform addition and subtraction. Instead, they should be able to explain the relationship between place value and base-10, particularly by writing multidigit numbers in terms of hundreds, tens, and ones, such as: 476 = (4 hundreds) + (7 tens) + (6 ones) and $476 = (4 \times 100) + (7 \times 10) + (6 \times 1)$.

Regrouping

At this stage, the standard algorithms for addition and subtraction are also introduced, and the concepts of "regrouping" (carrying or borrowing numbers across place values) are taught as part of direct instruction. However, for most children, it is often difficult to make a connection between writing multidigit numbers and adding/subtracting them horizontally (using expanded form) to vertically writing and adding/subtracting multidigit numbers by using carrying/borrowing (Ashlock, 2006). Research also shows that children can benefit from developing proficiency in the standard algorithms by learning to vertically add and subtract multidigit numbers by first using partial sums and partial differences (Figure 2.5) to help make better connections (Fuson et al., 1997).

Figure 2.5 Using Partial Sums to Develop Proficiency

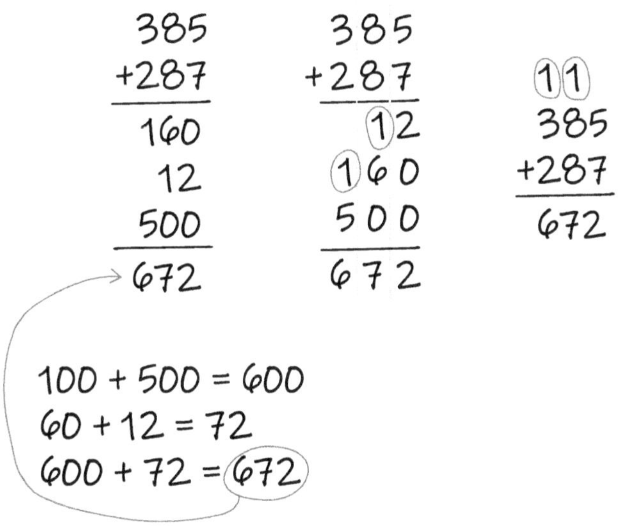

The first two solutions demonstrate how a student might use the partial sums method to regroup numbers, while the third solution uses the standard mathematical algorithm for addition.

Sidebar 2.7 Aligning Mathematical Expressions in Written Form on the Braillewriter

While Figure 2.5 shows the progression from partial sums to standard algorithms in print, a great way to help students practice spacing on the braillewriter is to use graph paper and number stickers, both available through APH (aph.org/product/embossed-graph-sheets-1-inch-squares and aph.org/product/feel-n-peel-stickers-nemeth-braille-print-numbers-0-100, respectively). The TSVI will have students line their numbers on the graph paper according to the problem, putting stickers of each number in the proper graph paper box. All blank boxes represent spaces that students will need to make with the spacebar on their braillewriter. This gives students a tactile representation of the problem that they then will transcribe on their own on their braillewriter.

The figure below shows how this method can be completed using graph paper, printed numbers, and sim braille. (For students with visual impairments, it is best to use tactile graph paper.) This is also a great way for TSVIs and transcribers of braille to ensure that the materials are properly transcribed for these students. Because the braille stickers also have print on them, this is an activity that the entire class can do together to practice aligning problems properly.

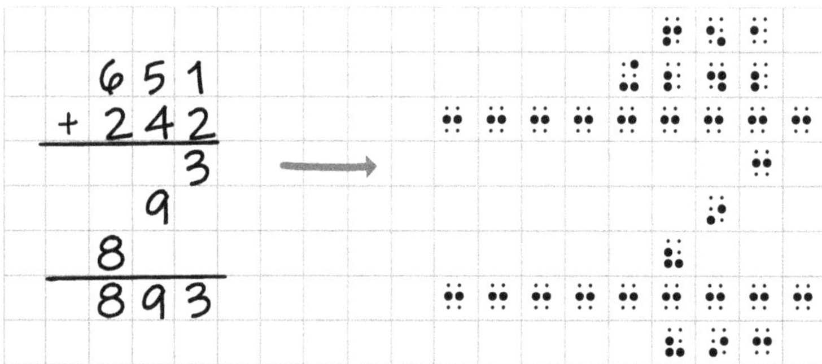

Additional resources on how these expressions are written can be found on the Paths to Literacy website (pathstoliteracy.org/resource/learning-and-teaching-the-nemeth-code).

The benefit of teaching partial sums and partial differences is that both are based on the conceptual understanding of place value, rather than algorithmic or procedural understanding, allowing students to add/subtract place values in any order they want and build upon their expanded form skills; with this method, they can check their answers, catch computation errors, and build confidence as they understand that difficult problems can be broken down into easier pieces. Students can also use the partial sums method to gradually develop the necessary skills of mathematical progressions to become more proficient in the standard addition algorithm, as shown in Figure 2.5.

Partial Sums

The student progressions in Figure 2.5 demonstrate developing proficiency in adding multidigit numbers vertically using expanded form and place values. This is a crucial foundational knowledge in a child's understanding of addition/subtraction of base-10 whole numbers. Within multidigit addition, discussion about partial sums will allow students to deeper understand place value. This is the method where students add the ones column first, regroup ("carry") a 1 (if needed), add all the 10s, and so on. This method is particularly accessible for students with visual impairments who are using UEB Math/Science or Nemeth code, as these codes focus on "parts" and then combine them at the end.

The partial sums method empowers many students to not only bridge the understanding between expanded form/place value; it also offers an opportunity for students who struggle with the standard algorithm to use another method that they may find possibly more intuitive. It allows them to gain proficiency and success in adding multidigit numbers and continue making progress within their grade-level expectations and mathematics curriculum as they continue to work on mastering the standard addition algorithm. As such, we strongly encourage TSVIs to initially focus on students and their needs to learn mathematics. Students should be taught to read spatially arranged problems using the exact braille formatting rules. However, the TSVI may need to collaborate with the general education teacher to ensure that the student masters solving these problems using an appropriate algorithm.

Subtraction

In regard to teaching the standard subtraction algorithm, research shows that this algorithm is very confusing to many children, leading to the development of math anxiety and frustrations, particularly for two cases: (1) if required to regroup or "borrow" and, more specifically, (2) if required to regroup or "borrow" from and across zeros (Ashlock, 2006). Specifically, many students who are proficient in subtraction with base-10 blocks still fall back on their "intuitive" reasoning rather than quantitative reasoning when required to regroup or "borrow" and to subtract the lower number from the higher number. In cases that require students to subtract/borrow across zeros (e.g., 4007 – 568), students often mistakenly borrow twice from the thousands place for each place value. Hence, it is important that teachers do not rush into teaching the standard subtraction algorithm but instead continue offering students more accessible (and less confusing) learning opportunities. Ideally, these opportunities should connect to their understanding of expanded form/place value and help make better sense of all subtraction cases conceptually, showing understanding of regrouping/borrowing, before moving to procedural/algorithmic understanding of subtraction.

Later, students will use what they have learned about place value to complete more advanced operations on whole numbers. Both multiplication and division build upon children's understanding of place value across all operations, including repeated addition and subtraction for multiplication and division.

Subtracting Using Number Lines and Vertical Addition

The *number line* (Figure 2.6) is one of the most familiar "tools" that students use and recognize when working with quantities. There are numerous resources involving number lines specifically developed for children with visual impairments; APH has dozens available (aph.org/search-results/?fwp_search_term=number+line). Number lines can easily be made accessible for students with visual impairments by using a tactile line and number stickers.

Teachers often use the number line to teach addition, yet rarely to teach subtraction. However, the number line helps to demonstrate the concept of subtraction as a (fixed) distance between two numbers (i.e., comparison). It establishes the fact that the distance between two numbers (such as 9 and 4) is the same as the distance between other sets of numbers (such as 10 and 5). Students discover that they can move this "fixed" distance left/right anywhere on the number line until they reach a set of two numbers that are easier for them to subtract.

Figure 2.6 The Number Line Method

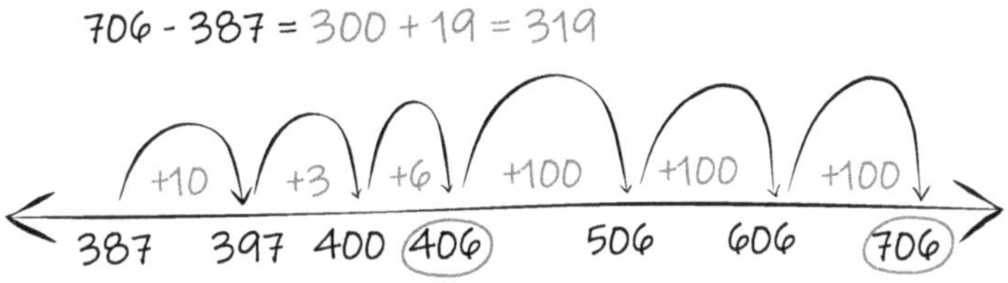

Number lines can be used for addition and subtraction.

Teaching the number line subtraction method (Figure 2.6) also can involve students reasoning about the answer being the distance between minuend (the number from which the other is subtracted) and subtrahend (the number subtracted)—for example: "How far is 387 from 706?" Teachers should allow students to share different ideas of how they might find that distance, as there are many ways to do so. Figure 2.6 shows a specific method that uses 406 as a *target number*: 406 is close to the subtrahend of 387, mirroring the final two digits of the minuend of 706. As such, it is easier to find the distance between 406 and 706, and students can focus on finding the distance between 387 and 406 first, then "hop" across the number line from 406 to 706. Number line "hops" are a great tool to depict relationships between numbers, including the introduction of *zero as the origin* or the starting number (to hop from) on the number line.

Figure 2.7 demonstrates the same reasoning by using *vertical addition*. In this method, students add numbers to reach toward the subtrahend until it becomes easy to add by 10s or 100s; then, they add the resulting numbers together to arrive at the answer. Both methods are very useful for helping students make connections between addition and subtraction by understanding the difference as a distance between two numbers. For students with low vision, they may be able to use both of the subtraction algorithms highlighted in Figures 2.6 and 2.7. However, for students who use braille, the TSVI and math teacher may need to develop different strategies. For example, a braille reader can use an accessible number line, but they may have to write the "movements" with their brailler instead. In many ways, doing this will lead them to Figure 2.7. While students with visual impairments should know how to read and spatially arrange problems using exact braille specifications, TSVIs and their students can develop alternative methods for solving problems within the classroom as needed.

Figure 2.7 Vertical Addition

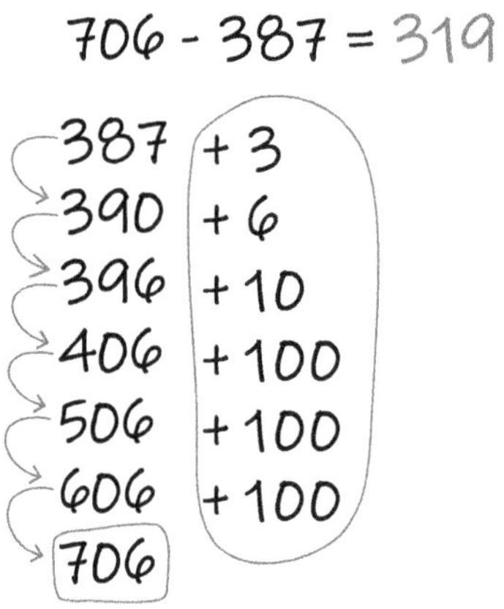

Vertical addition allows students to easily add 10s and 100s to reach the subtrahend.

Ultimately, when developing multidigit subtraction, we encourage teachers to recognize the complexity and confusing nature of the standard subtraction algorithm and offer students ample opportunities to use base-10 blocks for subtraction (including subtraction from and across zeros, such as 4007 – 568). They should also explore alternative methods and strategies that provide more intuitive learning and allow better conceptual connections—including some methods that do not require vertical subtraction or borrowing—to help children succeed as they continue making progress in mathematics.

USING (REPEATED) ADDITION AND SUBTRACTION TO DEVELOP MULTIPLICATION AND DIVISION SKILLS

Foundational knowledge is just that—a foundation upon which to build further expertise in progressively more complex operations. In K–2 grades, students engage in *skip counting* (counting by a number other than 1, e.g., 2, 4, 6, 8, etc.), which is very useful to building early multiplication skills.

As with addition/subtraction, the CGI project documents three levels of mathematical thinking and problem-solving strategies that students utilize to develop mastery for multiplication and division:

- **Level 1:** Direct modeling, concrete manipulatives, and one-by-one counting
- **Level 2:** Skip counting, repeated addition, multiplication strategies (e.g., doubles and halves, double and double again, double plus one more group, etc.)
- **Level 3:** Mental math strategies, multiplicative reasoning, derived facts

The story problems that students encounter involve two structures: *equal groups* and *multiplicative comparison problems* (Carpenter, Fennema, & Franke, 1996; Fuson & Fuson, 1992).

Equal groups multiplication and division problems involve students reasoning about a specific number of objects arranged equally into groups. A student's thinking and problem-solving strategies will change depending on whether the number of groups, number of items in a group, or total result is unknown. A basic equal groups problem that provides the number of groups and how many objects exist per group (e.g., "Jayden has 4 bags, and each bag has 6 toys. How many toys are there?") requires students to engage in *multiplication strategies* or *repeated addition.*

Multiplicative comparison problems ask students to compare how many times larger or smaller two quantities are (e.g., "The squirrel ate 2 pieces of bread, but the hungry duck ate 5 times as much bread. How many pieces of bread did the hungry duck eat?"). These types of problems engage students in skip counting, repeated addition, and multiplication strategies.

Division story problems mirror the above structures and ask students to employ different strategies across levels to arrive at the answer. For example, *fair share* division problems ask students to either find the size of each group, the number of groups, or total number of objects. If the size of each group is unknown (e.g., "Jayden has 24 toys and wants to share them equally between him and 3 friends. How many toys will each person receive?"), students might find the size of each group by engaging in *repeated subtraction*, backwards skip counting, or other mental math strategies.

Teaching Students With Visual Impairments Multiplication and Division

As with addition and subtraction, multiplication and division are best taught using tangible objects. Giving students objects that are already in bags and asking them to find the total is a great way to address multiplication, as it creates a concrete example of an equal groups problem. TSVIs should guide the students to explore the number of objects in each bag (using a tray will help ensure the objects are not lost) and then think back at how they would find the total number of objects. TSVIs can easily adapt this for division: For a fair share division problem, the students receive a total number of objects and must decide how many objects should be sorted into each bag for equal parts.

Students can also accomplish skip counting and repeated addition using a number line. A student can place a token or a counter at the first number in the equation as a placeholder on the number line, then use their finger to find the second number and mark it with a token or counter. They can then run their fingers along the number line to determine how many numbers lie between the first and last counter. From there, they may be asked to find an equal number of "hops" between the two numbers that they marked with counters, and teachers should encourage them to find as many different methods as possible. This will teach them to develop skip counting and repeated subtraction skills.

To determine mastery at Level 3 (mental math strategies), teachers should ask students to verbalize the procedures they are using to find their answers. This will ensure that teachers can informally assess their students' knowledge and problem solving when solving mathematical equations.

Use of Language in Multiplication and Division

The *compare multiplication* story problems often include the word "times" in the story, along with two categories of quantities to relate multiplicatively. For example, "Jill picked 6 apples, and Mark picked 4 times as many apples as Jill. How many apples did Mark pick?" (Van de Walle, Karp, Lovin, & Bay-Williams, 2013, p. 79). Similarly, the *compare division* problems also involve the word "times" and ask students to recognize the multiplicative relationship between two quantities or categories. For example, "Mark picked 24 apples. He picked 4 times as many apples as Jill. How many apples did Jill pick?" (Van de Walle, Karp, Lovin, & Bay-Williams, 2013, p. 79).

The goal of these structures is to move students (and teachers) away from the "keyword search" by supporting children's abilities to make sense of each problem: The problems both use "times," but they involve multiplication and division, respectively. Exposing students to this kind of nuance helps them develop different productive thinking strategies that lead to solution paths;

these may involve drawings, diagrams, tables, and using manipulatives. Note that story problems should begin with single-digit numbers before involving multidigit multiplication or division; this is because students must develop conceptual understanding of multiplication and division before moving to performing multiplication and division computations with multidigit numbers.

The Area and Partial Products Multiplication Methods

Effective teaching of multidigit multiplication and division involves several inherent challenges, primarily because teachers often rush to introduce the standard algorithms too early. It is important to introduce alternative methods for multiplication and division that have been shown to produce less chance of error and higher student success while continuing progress toward multidigit multiplication and division.

Two alternatives that students connect to extremely well are the *area* and *partial products* methods. In the area method, students create a divided rectangle that contains the numbers they wish to multiply broken down into their place values along the vertical and horizontal lines (see Figure 2.8). Students then multiply the numbers, effectively providing the "area" of each rectangle (though they will usually not be drawn to scale), then add the results together. In the partial products method, students multiply each digit of the larger number by each digit of the smaller number and add the results (see Figure 2.9). Both methods have straightforward organizational formats and mirror experiences students have had with *expanded form* methods of solving problems.

Figure 2.8 The Area Method

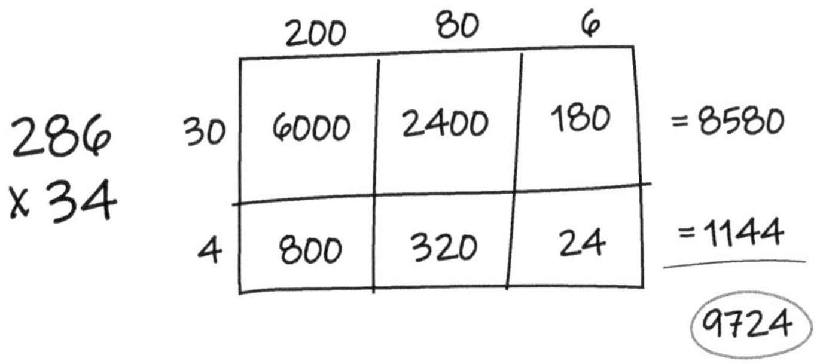

The area method involves multiplying numbers broken down by place value.

Figure 2.9 The Partial Products Method

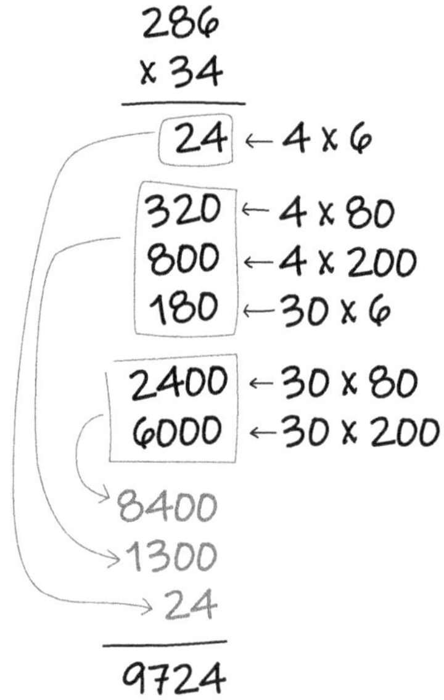

The partial products method involves multiplying each digit of the larger and smaller numbers together, then adding the results.

Teachers using area and partial products often share that their classroom artifacts and student work show conceptual understanding of multiplication, and students often calculate each (partial) product correctly. With these methods, most student errors actually stem from "other" mistakes, such as writing too many or not enough zeros (when multiplying the hundreds place), not keeping track of all the partial products, and/or adding all the "pieces" inaccurately.

There are adaptations that may help teachers better facilitate student progress and success with these multiplication methods: In the area method, students may be asked to color in zeros to ensure that the total number in each partial problem matches (in the top left-hand box of Figure 2.8, for example, 30 × 200 has three zeros, so there should still be three zeros in the result), add the results of each row separately to reduce complexity, or draw lines between place values to isolate the numbers being added. In the partial products method, teachers may ask students to write out the partial equation next to each line, group similar place values together when adding the results, or draw lines between place values.

Adapting the Area Method for Students With Visual Impairments

Both the area method and the partial products method can be made accessible with the use of manipulatives and other adaptations. For example, when using the area method, the TSVI can provide each number in the problem in braille on a separate card, which then can be put together to create the boxes found in the print above. Then, students can indicate the zeros in each row verbally and use tactile markers such as Wikki Stix pieces to place on the zeros in each row. On a separate piece of paper, students can then complete the mathematical problem or verbally provide the answer to each row. If students need to align the problem into proper notation, students can use number cards with technical code and align them vertically to achieve proper alignment in the problem. Tactually manipulating the cards into alignment creates the same effect as drawing the lines shown in the picture above. The TSVI may consider using the APH Brannan Cubarithm Slate (aph.org/product/brannan-cubarithm-slate) or the Math Window Basic Math Kit (mathwindow.com/basic-math) as a manipulative to support this method.

The Partial Quotients Division Method

Multidigit division is very difficult for Grade 4–5 students. Children especially struggle with the standard (long division) algorithm, which requires full proficiency in multiplication, as many fourth and fifth graders are still developing their multiplication skills. There is a more flexible and accessible method for students: the *partial quotients* method, which uses repeated subtraction of *easy groups* of the divisor, which students are more proficient with at this stage. (Easy groups are multiples of the divisor that are easy for students to create, such as 10 or 100, and that are smaller than the dividend.) For example, for 4,583 ÷ 27, students can think of methods to easily subtract easy groups of 27 from 4,583. Teachers can use number talks to prompt students to explain their ideas on how they might do that, helping them to develop specific thinking strategies that encourage students to think about easy groups and large easy groups of 27 that can be subtracted from 4,583 (e.g., 27 × 100 = 2,700). Teachers should also encourage students to reason about "doubles" and "halves" of these large easy groups (e.g., half of 2,700 is half of 100 groups, so 1,350 = 27 × 50). Figure 2.10 shows how a student can use easy large groups and implement thinking about halves into solving a problem via the partial quotients method.

Figure 2.10 The Partial Quotients Method

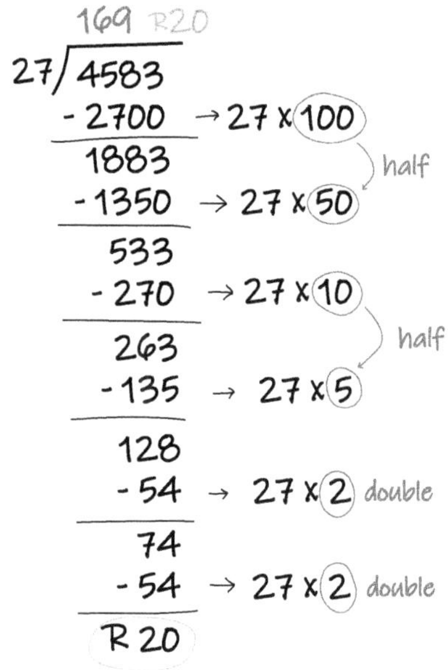

The partial quotients method relies on using repeated subtraction of easy groups to break down difficult division problems.

Research shows that many children who struggle with division exhibit more confidence with multidigit division as they continue subtracting easy groups, making noticeable progress in understanding division as repeated subtraction and by recognizing that the partial quotients method is a flexible process of subtracting groups that are easy for each child (individually) rather than everyone trying to find and subtract the largest group, as with long division (Hickendorff, Torbeyns, & Verschaffel, 2019).

Adapting Instruction to Student-Preferred Methods

We strongly encourage teachers to refocus their teaching more on children's preferred methods and their strengths in using these methods, rather than teaching the standard algorithm. Bouck, Long, and Jakubow (2022) found that the above alternative methods take less time for children to understand and yield higher percentages of success among children. Children who are able to master the partial products methods seem to exhibit less difficulty with making direct connections to the standard multiplication algorithm, especially when the two methods (partial products and standard algorithm) are written side by side and used, via number talks, to prompt students to notice and describe similarities and differences using connections between the two methods. It is important to note that explanations for how the "carry" numbers in the standard algorithm also appear in the partial products method (Ambrose, Baek, & Carpenter, 2013).

In fact, research shows that children who are proficient in multiplication still struggle with division, particularly with finding the largest group, and spend considerable time and work in the trial-and-error process, testing and continually writing the multiplications on the side (to check whether the group(s) are indeed the largest (Hickendorff, Torbeyns, & Verschaffel, 2019). Conversely, children who have become comfortable with repeated addition and the partial quotients method take less time to solve mathematical problems, their work yields a smaller chance of error, and the method allows them to use their work to check and correct their answers/mistakes (Ambrose, Baek, & Carpenter, 2013).

After introducing alternative methods to young students and providing them with enough time to become more comfortable, it is important that teachers continue to implement number talks to teach numbers and operations, including multidigit numbers and operations. Number talks offer children, especially lower achieving and struggling students, opportunities to engage and share and learn strategies among their peers. This has particular benefit for students who are blind or visually impaired, as having the opportunity to talk about their perceived experiences with numbers increases their understanding of mathematical concepts (Ahlberg and Csocsán, 1999).

CONCLUSION

The elementary grades are a critical time for learning mathematics, and students who are visually impaired may underperform relative to their peers during this period. However, according to Ahlberg and Csocsán (1999), "Although there is a gap between the development of blind children's and sighted children's arithmetical competence in the early years, this gap disappears by the time the child reaches 8–11" (p. 549). That gap can only close with structured, intentional experiences that provide students with a strong foundation in whole-number knowledge. Teachers must consistently provide accessible, tactual experiences, including opportunities to work on a braillewriter, accompanied by precise mathematical language. A considered approach that helps students develop conceptual mastery will prepare them for more complex computations in the later grades.

REFERENCES

Ahlberg A., & Csocsán, E. (1999). How children who are blind experience numbers. *Journal of Visual Impairment & Blindness, 93*(9), 549–560. https://doi.org/10.1177/0145482X9909300902

Ambrose, R., Baek, J. M., & Carpenter, T. P. (2013). Children's invention of multidigit multiplication and division algorithms. In A. J. Baroody & A. Dowker (Eds.), *The development of arithmetic concepts and skills: Constructive adaptive expertise* (pp. 307–338). Routledge.

Antell, S. E., & Keating, D. P. (1983). Perception of numerical invariance in neonates. *Child Development, 54*(3), 695–701.

Ashlock, R. (2006). *Error patterns in computation: Using error patterns to improve instruction*. Pearson/Merrill Prentice Hall.

Bisanz, J., Sherman, J. L., Rasmussen, C., & Ho, E. (2005). Development of arithmetic skills and knowledge in preschool children. In J. I. D. Campbell (Ed.), *The handbook of mathematical cognition* (pp. 143–162). Psychology Press.

Bouck, E. C., Long, H., & Jakubow, L. (2022). Examining an intervention package to teach students with mathematical difficulties partial products online. *Journal of Computers in Mathematics and Science Teaching, 41*(1), 5–27. https://doi.org/10.70725/166270qrfmtx

Butler, D. L., Beckingham, B., & Lauscher, H. J. N. (2005). Promoting strategic learning by eighth-grade students struggling in mathematics: A report of three case studies. *Learning Disabilities Research & Practice, 20*(3), 156–174. https://doi.org/10.1111/j.1540-5826.2005.00130.x

Carpenter, T. P., & Fennema, E. (1992). Cognitively guided instruction: Building on the knowledge of students and teachers. *International Journal of Educational Research, 17*(5), 457–470. https://doi.org/10.1016/S0883-0355(05)80005-9

Carpenter, T. P., Fennema, E., & Franke, M. L. (1996). Cognitively guided instruction: A knowledge base for reform in primary mathematics instruction. *The Elementary School Journal, 97*(1), 3–20. https://doi.org/10.1086/461846

Carpenter, T. P., Fennema, E., Franke, M. L., Levi, L., & Empson, S. B. (1999). *Children's mathematics: Cognitively guided instruction* (Vol. 1). Heinemann.

Common Core State Standards Initiative (CCSSI). (2010). *Common core state standards for mathematical practice.* National Governors Association Center for Best Practices and Council of Chief State School Officers (CCSSO).

Fuson, K. C., & Fuson, A. M. (1992). Instruction supporting children's counting on for addition and counting up for subtraction. *Journal for Research in Mathematics Education, 23*(1), 72–78.

Fuson, K. C., Wearne, D., Hiebert, J. C., Murray, H. G., Human, P. G., Olivier, A. I., Carpenter, T. P, & Fennema, E. (1997). Children's conceptual structures for multidigit numbers and methods of multidigit addition and subtraction. *Journal for Research in Mathematics Education, 28*(2), 130–162. https://doi.org/10.2307/749759

Gallistel, C. R., & Gelman, R. (1992). Preverbal and verbal counting and computation. *Cognition, 44*(1–2), 43–74. https://doi.org/10.1016/0010-0277(92)90050-R

Gelman, R. (1980). What young children know about numbers. *Educational Psychologist, 15*(1), 54–68. https://doi.org/10.1080/00461528009529216

Gelman, R., & Gallistel, C. R. (1978). *The child's understanding of numbers.* Harvard University Press.

Gelman, R., & Tucker, M. F. (1975). Further investigations of the young child's conception of number. *Child Development,* 167–175. https://doi.org/10.2307/1128845

Ginsburg, H. P., & Seo, K. H. (1999). Mathematics in children's thinking. *Mathematical Thinking and Learning, 1*(2), 113–129. https://doi.org/10.1207/s15327833mtl0102_2

Hickendorff, M., Torbeyns, J., & Verschaffel, L. (2019). Multi-digit addition, subtraction, multiplication, and division strategies. In A. Fritz, V. G. Haase, & P. Räsänen (Eds.), *International handbook of mathematical learning difficulties: From the laboratory to the classroom* (pp. 543–560). Springer.

Ivy, S., & Hooper, J. (2015). Using constant time delay to teach braille and the Nemeth code for mathematics and science notation to students making the transition from print to braille. *Journal of Visual Impairment & Blindness, 109*(5), 343–358. https://doi.org/10.1177/0145482X1410800203

Maccini, P., & Hughes, C. A. (1997). Mathematics interventions for adolescents with learning disabilities. *Learning Disabilities Research & Practice, 12*(3), 168–176.

Moscardini, L. (2014). Developing equitable elementary mathematics classrooms through teachers learning about children's mathematical thinking:

Cognitively guided instruction as an inclusive pedagogy. *Teaching and Teacher Education*, *43*, 69–79. https://doi.org/10.1016/j.tate.2014.06.003

National Center on Intensive Intervention at the American Institute for Research. (n.d.). https://intensiveintervention.org

National Research Council, Division of Behavioral and Social Sciences and Education Center for Education, & Mathematics Learning Study Committee. (2001). In B. Findell, J. Swafford, & J. Kilpatrick (Eds.), *Adding it up: Helping children learn mathematics*. National Academies Press.

Parrish, S. D. (2011). Number talks build numerical reasoning. *Teaching Children Mathematics*, *18*(3), 198–206. https://doi.org/10.5951/teacchilmath.18.3.0198

Phan, T., & Conroy, P. (2022). Counting skills of young children with visual impairments. *Journal of Blindness Innovation and Research*, *12*(1). https://doi.org/10.5241/12-230

Piaget, J. (1952). *The child's conception of number*. Norton.

Rottmann, T., Haberzettl, N. & Kramer, M. (2020). Inclusive assessment of whole number knowledge—development and evaluation of an assessment interview for children with visual impairments in the primary grades. *Mathematics Education Research Journal*, *32*, 147–170. https://doi.org/10.1007/s13394-019-00296-9

Smith, D. W., & Smothers, S. M. (2012). The role of characteristics of tactile graphics in secondary mathematics and science textbooks in braille. *Journal of Visual Impairment & Blindness*, *106*(9), 543–554. https://doi.org/10.1177/0145482X1210600905

Tournaki, N., Bae, Y. S., & Kerekes, J. (2008). Rekenrek: A manipulative used to teach addition and subtraction to students with learning disabilities. *Learning Disabilities: A Contemporary Journal*, *6*(2), 41–59.

Van de Walle, J. A., Karp, K. S., Lovin, L. H., & Bay-Williams, J. M. (2013). *Teaching student-centered mathematics: Developmentally appropriate instruction for grades 3–5* (Volume II). Pearson.

Vygotsky, L. S. (1978). *Mind in society: The development of higher psychological processes*. (M. Cole, V. John-Steiner, S. Scribner, & E. Souberman, Eds.) Harvard University Press.

Wynn, K. (1992). Children's acquisition of the number words and the counting system. *Cognitive Psychology*, *24*(2), 220–251. https://doi.org/10.1016/0010-0285(92)90008-P

CHAPTER 3

EXPANSION OF NUMBERS AND OPERATIONS: FRACTIONS AND DECIMALS

Aina Appova and Tiffany Wild

INTRODUCTION

The expansion of students' foundational knowledge to develop a conceptual understanding of fractions and operations on fractions requires students to have a strong number sense, understanding of place value, and geometric reasoning to be able to partition a region into equal (fractional) parts/segments. Students, particularly elementary school students, perceive fractions as one of the most difficult mathematics topics to learn due to the seemingly counterintuitive notations used and produced. The National Mathematics Advisory Panel (NMAP, 2008) reported that more than 50% of U.S. middle and high school students, including students with visual impairments, struggle with elementary-level fraction content.

MISCONCEPTIONS WHEN LEARNING ABOUT FRACTIONS

One of the most common areas of misconception for students is how to treat numerators and denominators when adding fractions (Ashlock, 2006). Traditional teaching practices, which focus heavily on following the teacher's steps, procedures, and memorizations, can negatively impact student learning and mathematical understanding (Misquitta, 2011). Studies examining the pedagogy used to teach fractions to students with visual impairments and disabilities show that these traditional teaching practices are prevalent (Hunt et al., 2023). These pedagogical methods lead to confusion and misconceptions that persist into the later grades and can cause mathematics anxiety, low self-esteem, and declined motivation to learn for all types of students, including those with visual impairments (Behr, Wachsmuth, Post, & Lesh, 1984; NMAP, 2008). Greater difficulties were particularly observed in the populations of low-achieving and historically underserved students, including students with disabilities, as these populations are traditionally lagging at least two grade levels behind, particularly in fractions (NMAP, 2008; Blackorby & Wagner, 1996). U.S. data also shows that teachers continue to teach fractions using traditional and ineffective pedagogies (such as show-and-tell and memorization), and students continue to struggle with fractions and carry their confusions and misconceptions into later grades (Behr, Wachsmuth, Post, & Lesh, 1984; NMAP, 2008).

In this chapter, we strongly emphasize students' opportunities to learn, particularly for *conceptual understanding* as a path to procedural fluency, not the other way around. We urge teachers to follow one specific teaching goal: to design and provide students with learning opportunities that focus on exploring fractions *intuitively and conceptually* by building on what students already know and understand about numbers—before introducing any symbolism or procedures. This approach will better support students' learning and mathematical development. This chapter provides specific activities, questions, and classroom tips and suggestions to help teachers achieve this goal. Data shows that students with strong conceptual understanding of fractions make much deeper and more meaningful connections that also help them bridge the gap between conceptual and procedural understanding of fractions, including developing their own methods and procedures to correctly perform operations on fractions (Hiebert & Grouws, 2007; McLeod & Armstrong, 1982; Cai & Ni, 2011).

DEVELOPING STUDENTS' UNDERSTANDING OF FRACTIONS

Teacher Concerns About Teaching Fractions

Most teachers, particularly those teaching elementary math, share students' trepidation toward fractions, reporting that they are too abstract for students (Ball, Hill, & Bass, 2005; Hiebert & Grouws, 2007; Mack, 2001). For teachers of students with visual impairments (TSVIs), this attitude may have an outsize impact. A study reported that fractions were the second most frequently prepared operation TSVIs used Nemeth code for, preceded only by basic number operations such as addition, subtraction, multiplication, and division (Rosenblum & Amato, 2004). However, it is not enough to simply know the code: "They [TSVIs] must have an understanding of math themselves so that Nemeth code makes sense to them" (Rosenblum & Amato, 2004, p. 490). Whether it is Nemeth code or Unified English Braille (UEB) Math/Science, it is vital that the TSVI is comfortable with the underlying mathematical principles their students are exploring.

Most textbooks reference teaching "part–whole" relationships between the numerator and denominator as part of teaching fractions. Those same textbooks often use visual-based material to represent this relationship, such as by presenting the fraction as a square that has been divided into fourths with one shaded area. This method makes the concept inherently inaccessible to students with visual impairments (Emerson & Anderson, 2018). Even though evolving technology strives to lessen this inaccessibility, textbooks that include screen reader capabilities often cannot read fractions or their visual representations if they are coded incorrectly, referring to them only as "image."

In this chapter, we elaborate on the mathematical complexities that "part-whole" relationships entail, particularly the multifaceted nature of the mathematics involved and required of students to be successful with understanding these relationships. These complexities are twofold: There is the introduction of added symbolic representations and notations, and there is the conceptual knowledge and connections required across multiple areas of mathematics. This includes (but is not limited to) number sense, whole numbers and operations, geometry and spatial reasoning, measurement, discrete mathematics, and logical reasoning (particularly related to understanding equivalence relations). In order for students to be successful with fractions, they must use each of these areas of mathematics. Students must also remember that rational numbers—including whole numbers, negative numbers, and decimals—are all numbers that students can express as a fraction as long as they possess a nonzero denominator. For example, $\frac{4}{1}$ is equivalent to the whole number 4.

Early Introduction to Fractions

Students receive an introduction to fractions at a very early age, typically in the form of "equally" sharing a candy bar or a cookie with friends and being able to reason about the size of the fractional part that each friend would receive. This developing understanding of fractional parts of a whole cookie eventually and essentially lead to the required reasoning about *equal* fractional parts of the whole. Students mostly struggle with equal part-to-whole relationships in two areas: When the whole is no longer one continuous piece of candy and the *equal* part-to-whole relationships change with the increase in the number of friends (e.g., unit fractions of different size: 5 friends = $\frac{1}{5}$ of the candy bar per friend) or the number of cookies (which can lead to improper fractions: 10 cookies equally shared by 4 friends = $2\frac{1}{2}$ cookies per friend).

Teaching Students With Visual Impairments

For students with visual impairments, it is essential to begin using proper mathematical language when talking about equal fraction parts. For example, "I am breaking this cookie in half. Do you want half?" Showing the child the entire candy bar and then the parts of the candy bar that equal eighths or fourths is also beneficial. This becomes the basic introduction of fractions and manipulatives used in formal education.

Chapter 2, "Foundations of Numbers and Operations," mentions number talks as an instructional strategy that engages students in sharing their ideas and thinking strategies. As an extension of number talks, teachers should use story problems daily to help students who are learning fractions explore different fractional parts of the whole by using and connecting to proper mathematical language; this is particularly powerful for students with visual impairments.

For example, a great extension question would be asking students to think about dividing 6 cookies *equally* among 4 friends. In order to solve this problem, students may develop two solution strategies: (1) considering the "fair-share" strategy to equally divide 4 cookies and splitting the remaining 2 cookies in half, which result in 1 and $\frac{1}{2}$ cookie (per friend); and (2) splitting each (whole) cookie among 4 friends (4 equal parts, or $\frac{1}{4}$ cookie per friend), arriving at the answer of each friend receiving 6 pieces of $\frac{1}{4}$ cookie, or $\frac{6}{4}$ cookie ($6 \times \frac{1}{4} = \frac{6}{4}$). Both solutions are mathematically correct, valid, and appropriate, and these types of story problems expand students' understanding of the equal part-to-whole relationship, particularly in their ability to find both solutions, understand their mathematical differences, and recognize their equivalence.

Furthermore, these discussions (and story problems) offer students opportunities to make direct connections and extensions from their knowledge of division to fundamental concepts of fractions. Teachers should provide these opportunities to students as required initial experiences with fractions via pictures, manipulatives, story problems, and mathematical conversations.

Sidebar 3.1 Nemeth in a Box

A great resource to facilitate number talks is available on the Paths to Literacy website: Nemeth in a Box (pathstoliteracy.org/nemeth-box-middle-school-students). By using math games, students find encouragement not only to practice their written Nemeth code skills but also their problem-solving and reasoning skills. Each lesson includes a facilitator guide, a presentation for an overview of instruction, and student handouts. TSVIs should encourage students to provide verbal reasoning for why they chose different answers in the games.

For students with visual impairments, teachers should start with the actual representations being divided, such as providing each student 6 cookies (if they are engaging with the problem described above). The actual representations will allow students to tactually explore the "whole" cookie and tactually engage to find equal parts. Another great representation is a candy bar. Although messy, many chocolate bars have tactual markers on them that can help students understand how to break apart the pieces to find equal parts. (Tip: Refrigerating the bar will keep it from melting as fast, enabling the teacher to tactually break it into pieces.)

The class can slowly move from actual cookies or candy bars to fraction circle models or any models that provide representations of partitioning a whole into equal and inequal parts. One resource to explore *fractions via play* during early childhood ages is the Tactile Tangrams Kit available from the American Printing House for the Blind, or APH (aph.org/product/tactile-tangrams-kit), which helps students learn basic shapes and their textures, colors, and sizes, in addition to exploring "parts" of these shapes as an informal introduction to fractions. Once students have a mental model of the object, they can think through the problem.

Engaging with tactile representations of problems also allows the child the ability to engage in deeper understanding of the mathematical concepts as students learn the more formal concepts of parts and wholes. Informal experiences with fractions offer students opportunities to access this knowledge early and provide an opportunity to build deeper conceptual, tactile, and foundational knowledge for understanding fractions. Teaching fractions with manipulatives is discussed in depth later in this chapter.

Sidebar 3.2 Fraction Manipulatives

The APH Mathbuilders: Fractions kit (aph.org/product/mathbuilders-fractions) includes accessible fraction circles and fraction tiles. Additional resources for making homemade fraction circles are available from StarLitDay (starlitday.com/2021/02/make-your-own-fraction-circles-and-frames). Once the kit is assembled, the teacher presents portions of a circle in different fractional increments. The students then use the circle frames to create partial or whole circles using various part sizes.

Formal Introduction to Fractions

Formally, according to Common Core Standard 1.G (CCSSI, 2010), introduction of fractions should begin as early as Grades K–5 (5–10 years old). This standard requires students to

> Partition circles and rectangles into two and four equal shares, describe the shares using the words *halves*, *fourths*, and *quarters*, and use the phrases *half of*, *fourth of*, and *quarter of*. Describe the whole as *two of*, or *four of*, the shares.

By Grade 3, students must (1) understand a fraction $1/b$ as the quantity formed by 1 part partitioning a whole into b equal parts; (2) understand a fraction a/b as the quantity formed by a parts of size $1/b$; (3) understand and represent a fraction as a number on the number line; and (4) explain the equivalence of fractions and compare fractions by reasoning about their size (3.NF; CCSSI, 2010). So, in the span of two years, teachers must provide students with learning opportunities that develop their required skills and proficiencies.

The aforementioned first-grade standard requires students to reason *geometrically* and *spatially*; it is a *geometry* standard, after all. Students need to build upon their understanding of area of polygonal (angular) and curved regions by partitioning these regions into equal parts, all while paying precise attention to congruent and noncongruent parts of equal areas and correctly identifying the size of each part in relation to the (whole) area of the region. First graders will not yet have access to the mathematical vocabulary; hence, teachers can address this standard via *visual and tactual* experiences that allow students to approach and discuss this concept using their knowledge of *congruence* (elements being of the same shape and identical size) of equal parts or pieces. TSVIs can incorporate APH's Tactile Tangrams (aph.org/product/tactile-tangrams-kit) or MathBuilders Fractions (aph.org/product/mathbuilders-fractions) to support these explorations.

Teaching Fractions Through Congruence

Partitioning

Teachers should introduce students to congruent parts while developing their early understanding of the meaning and language of "halves," "thirds," "fourths," and being able to partition the whole into those equal-sized parts or pieces. In turn, students should show understanding that the number of equal-sized pieces represent the correct fraction (e.g., "fourths" will result in a total of four equal-sized pieces). Teachers can help students relate to these representations via examples of splitting a cookie or cutting a pizza in half both vertically and horizontally to make fourths. Paper-folding activities are also very helpful in developing these concepts, as these activities draw on the child's kinesthetic and tactile learning through touching and folding papers and producing a multisensory experience involving mathematics.

For students with visual impairments, cardstock paper allows them to create distinct lines as they fold the paper and then count the number of equal partitions to determine the fractions. It is also important that teachers ask guided questions to allow the students to brainstorm ways to fold the paper into a new fraction. For example, after the students fold the paper in half, ask them how they think they could fold the paper to get four equal boxes and then identify the fraction $\frac{1}{4}$. Extension activities could also involve students folding paper to create additional equal parts (such as eighths), as the concept of "halving" the fourths can make for a powerful connection to the concept of eighths. First graders can access this knowledge tactually by paper folding and without performing fraction multiplication.

Again, it is important that teachers use these activities with carefully designed math questions that require students to reason about the size of each piece as the denominators increase or decrease; this allows students to develop an informal understanding of area and proportionality. Research shows that even when a student's understanding of *partitioning* is strong, it is partitioning into equal parts that is most difficult when learning fractions. Therefore, extensive kinesthetic and tactual experiences with this partitioning produce significant differences in students' learning outcomes regarding fractions (Ashlock, 2006; Mohyuddin & Khalil, 2016).

Additionally, at the initial introduction of fractions, teachers should provide students with kinesthetic and tactual learning opportunities to specifically identify the *equal parts* of polygons and circles, focusing on the *language* (e.g., halves, thirds, fourths) and on the *actions* of folding, cutting, drawing, and coloring the equal parts. Teachers also should provide students with counterexamples (ideally using students' work, including their errors) to discuss the congruent/noncongruent (equal-sized/unequal-sized) parts in reference to the whole.

Identifying Equal and Unequal Parts of Different Wholes

When teaching fractions, teachers must continuously ask students to define the *whole* and the *parts*, as they are among the most important (and difficult to master) learning goals for students. The difficulty students encounter is partly due to a mathematics curriculum that offers students limited experiences and understanding of fractions—particularly that the whole is always the entire solid shape and the part is always smaller than the whole. By Grade 2, teachers should include multiple activities that offer students specific explorations of fractions where the whole is not a static, fixed, or solid piece, and students should begin to understand that not all parts of a whole are equal.

The shapes presented in Figure 3.1 demonstrate how equal and unequal parts make up a whole. Teachers can use shapes such as these to help students build upon their knowledge of equal and unequal parts by asking them to define each inside shape as a fraction in terms of size relative to the whole and then name/write each fraction symbolically. These shapes also introduce the circle as a curved shape/region that students can also learn to partition into equal-sized parts and write/name the fractions.

Figure 3.1 Equal and Unequal Parts

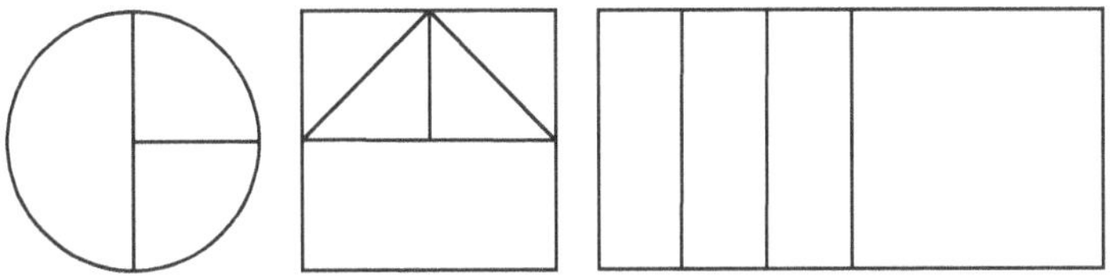

Exploring shapes divided into equal and unequal parts helps students understand the relationships between parts and wholes.

For example, when considering the five different shapes making up the middle square in Figure 3.1, students may state that the triangles are "eighths" and the large square is the "whole." This is a very good starting point that shows students' basic understanding of the part–whole relationship. However, teachers must build upon this understanding and use extension activities for students to find and define other possible fractions, parts, and wholes in the same picture. These extensions and discussions should require students to recognize and describe the shapes inside the large square as wholes as well and also understand how changing the whole affects the parts and the fraction.

For example, take the upper left square within the square of Figure 3.1 made up of two triangles. By examining the square, the concept of "whole" can shift: If the square is the "whole," then the triangles inside the square become "halves." Another way to examine parts would be to ask students how they could divide the bottom rectangle of Figure 3.1 in the same way as the top rectangle, necessitating four triangles in each half of the square. Eight triangles would then form the original square, making the original square eight "wholes." These discussions are critical and necessary to help strengthen students' understanding of part–whole relationships and development of mathematical flexibility when understanding part–whole relationships.

Teachers can then move students beyond the assumptions that basic shapes such as hexagons are always *whole* by asking them to consider the different shapes that make up that whole, such as a trapezoid, rhombus, or triangle (see Figure 3.2). A teacher can pose questions such as: "If the hexagon is the whole, what part of the whole is the triangle?" and "If the trapezoid is the whole, what part of the whole is the triangle?" and "If the rhombus is the whole, what part of the whole is the triangle?" (Please note: These questions specifically reference the whole that is still bigger than the parts, yet they also refocus students' attention on the same part—for example, a triangle—that becomes a different fraction depending on the whole under consideration.)

Figure 3.2 Demonstrating the Relationship Between Wholes and Parts

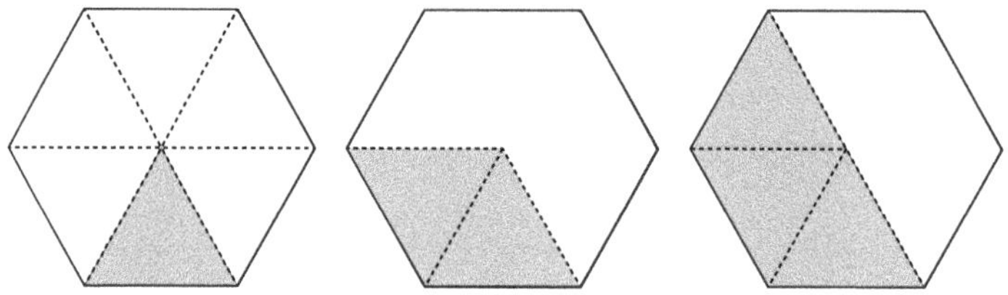

A hexagon can be divided into simpler shapes in several ways.

Additionally, teachers should pose questions in which the whole is smaller than the parts, such as: "If the trapezoid is the whole, what part of the whole is the hexagon?" and "If the rhombus is the whole, what part of the whole is the trapezoid?" and "What part of the rhombus is the hexagon?" These questions help students develop a deeper understanding of the part–whole relationship and introduce them to improper fractions tactually and more meaningfully.

Making Activities Concerning Equal and Unequal Parts Accessible

The above activity is present in many schools but can be difficult for students with visual impairments who are not easily able to study the shapes. For students with low vision, drawing the outline of shapes on a clear overlay, placing it onto a light box, and then providing students with shaped tiles may help them participate in the activity. Often, once the light box is in use, all students will want to use it for the lesson, and the lesson becomes accessible for all, normalizing the accommodation.

For students who are unable to see the shapes, this activity can be more challenging. To address this, place textures atop corresponding shapes—for example, the triangles that come together to form a square in Figure 3.1 could have a similar texture (such as felt or sponge), while the rectangle that makes up the remainder of the square could have a different texture. Outline the entire shape with a raised line (using Wikki Stix, yarn, etc.) to help the students understand the outer boundaries of the activity. Subsequently, if needed, each individual larger shape could feature a raised outline. For example, the upper left square could have an outline, and the students would attempt to find the two shapes that make the smaller square (i.e., the two textured triangles).

All students with visual impairments can benefit from tactual representations, which serve as specific examples that can help students make better sense of the differences between parts and equal parts, including references to halves, fourths, and thirds in relationship to the whole. The learning goal for students would be to discuss, explain, and recognize these partitions as equal parts to identify each part in relationship to the whole. At this stage, teachers should ask students to state the fraction for each of these parts and explicitly describe the relationship of the denominator to the size of the part.

Tessellations

Young students may best learn the concept of congruence through *tessellations* (i.e., repeatedly putting shapes together to cover a surface/plane without any gaps). To this end, teachers should focus students' explanations on the shapes' congruence. Encouraging students to describe the shapes using the words *equal size* and *equal shape* provides students opportunities to discuss noncongruent parts that come together to make the entire shape. Focus on using the terms *unequal parts*, *bigger/smaller*, or *uneven/unevenly covered area* of the entire (whole) original shape. For these activities, pattern blocks (see Figure 3.3) are very helpful to teach fractions, helping students make geometric connections and recognize relationships between smaller shapes that can come together to make larger shapes. Students can do this through the use of tessellations, tiling, or stacking. In the top image of Figure 3.3, a smaller trapezoid, parallelogram, and triangle are arranged to demonstrate that they make up a hexagon, and in the bottom image, the smaller shapes are grouped together around the perimeter of a hexagon to form a larger hexagon. Asking students to find these kinds of relationships between shapes requires them to use their skills of geometric reasoning to develop *fractional reasoning*, recognizing and describing how these polygons are halves, thirds, or sixths of the hexagon.

The tiles in Figure 3.3 are color coded, which is beneficial to students with low vision. However, all students can benefit from using the tactile exploration of pattern blocks to explain part-to-whole relationships. Giving these shapes different textures provides additional information for students with visual impairments to be able to more easily find similar shapes. For example, a glue dot placed on all the green triangles or a wavy glue line on all the blue rhombus pieces would not only encourage students with visual impairments to explore the shape tactually but also to quickly sort shapes, which their peers would be doing visually.

Figure 3.3 Demonstrating How Different Shapes Make a Whole

Using manipulatives to explore tessellations will help students understand the relationship between shapes.

Pattern blocks are ideal for developing mathematical flexibility that may be harder to grasp from partitioning diagrams like those in Figure 3.1. Teachers should encourage students to continue to make different larger shapes by putting together smaller geometric shapes, all while answering probing questions regarding their understanding of part-to-whole concepts to promote student understanding. For example, after asking a student to re-create the lower image in Figure 3.3, a teacher might ask the following questions:

- How many triangles would make up the yellow hexagon? [Answer: 6.]
- How about the red trapezoid? [Answer: 3.]
- How about the blue rhombus? [Answer: 2.]
- How many shapes are in the entire hexagon? [Answer: 10.]
- So, using numbers, how would we represent the triangles as part of the entire hexagon? [Answer: 3 triangles as part of a total of 10 pieces; however, 10 pieces comprise 24 triangles in the entire hexagon. Hence, the triangles represent $\frac{3}{24}$.]

Measurement/Length/Height Model

The *measurement* model features the whole as one continuous piece/object/region (e.g., hexagon, trapezoid, pizza/circle, brownie/rectangle, etc.). Measurement is an area of mathematics used in many contexts that are very important for students with visual impairments to understand (Smith, 2017). Measurements used to teach fractions allow students to reference the *length* of the whole, rather than the area, and TSVIs should provide many opportunities for students with visual impairments to practice measurement in a variety of mathematical contexts.

Some teachers default to using the number line to teach fractions as the measurement model, but—at least initially—this limits the mathematical understanding and classroom discussions of what happens to the same parts when the whole changes. This is because, on the number line, the whole does not change; changing the whole on the number line is not mathematically appropriate. Hence, students may enjoy better use of the line for benchmarking the whole by halves, fourths, thirds, etc., and additional appropriate manipulatives to teach the measurement model include Cuisenaire rods, fraction tower cubes, or fraction tiles. These manipulatives allow students to "measure" the size of the fraction using length/distance (or height, if stacking) and experience the whole as not just a circle/rectangle (e.g., pizza or brownie) but also as a *part–whole* relationship between different lengths (parts) as they relate to the entire specified distance (whole). Teachers should introduce measurement manipulatives that do not come prelabeled to allow students to build deeper, richer, and stronger mathematical conceptions of fractions.

Manipulatives in Teaching Within the Measurement Model: Cuisenaire Rods, Fraction Tower Cubes, and Fraction Tiles

Cuisenaire rods are an ideal tactile manipulative that, like pattern blocks, do not come prelabeled. This allows students to conceive of any rod as the whole and find all the different *part–whole* relationships to that rod/whole.

When teaching concepts with the Cuisenaire rods, teachers should continue asking students to identify a rod that represents a half of another rod or a nonunit fraction, such as: "If the tallest rod created is the whole, what fraction is a smaller rod to the taller rod?" The teacher can then use different lengths of rods to have students compare a part to the whole. Allow students time to explore the different lengths of the rods and line them up side by side for comparison. For students with low vision, teachers should refer to the colors of each rod they would like students to compare; for other students with visual impairments, use tactile indicators or braille labels (as discussed with pattern blocks previously) on each of the rods.

Figure 3.4 Cuisenaire Rods

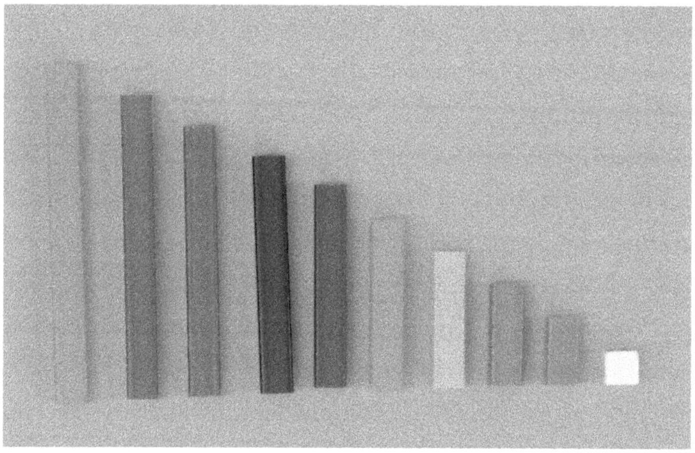

Cuisenaire rods allow students to explore part–whole relationships.

These types of questions also introduce students to groups of unit fractions (e.g., if 3 red rods make a whole, $\frac{2}{3}$ = 2 red rods; also $2 \times \frac{1}{3}$ = two groups of one third). To help students develop conceptual understanding of unit and nonunit fractions, it is essential to pose deeper-level questions, such as: "If the dark green rod is a whole, which rod is two thirds?" and "If the light green is one third, what rod is the whole?" These types of questions not only help develop students' understanding of fractions but also use fraction-based language to help students become more precise in their reasoning.

Another method that can benefit students in actually seeing the fractions that make up a part–whole relationship is the use of fraction tiles and fraction cubes, as shown in Figure 3.5. These objects display the numbers of the whole and the parts. Students can easily move the pieces to show how, for example, 4 smaller tiles or blocks can combine to make a whole. Students can then see that $\frac{1}{4} + \frac{1}{4} + \frac{1}{4} + \frac{1}{4} = 1$, or 1 whole. When using these materials, teachers should place braille labels over the numerals and fractions in either UEB Math/Science or Nemeth code. This allows students to not only understand part–to–whole concepts but also be able to students to practice reading and writing each fraction properly. The MathBuilders Fractions kit (aph.org/product/mathbuilders-fractions), includes similar ready-made tiles and fraction blocks.

Figure 3.5 Fraction Tiles

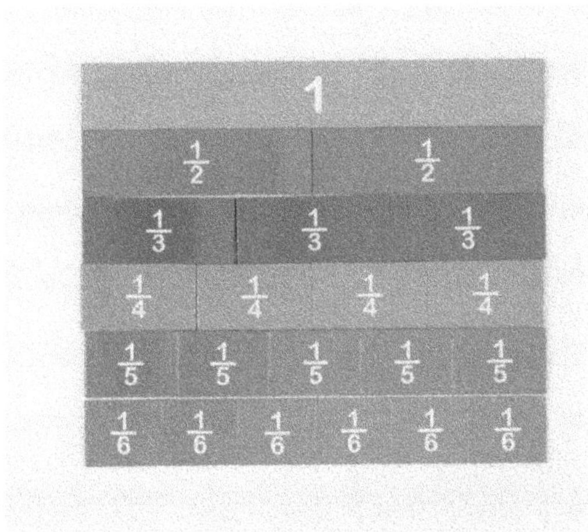

Fraction tiles (and other fraction manipulatives) predefine the parts and wholes and demonstrate the relationship between fractions.

It is important to remember that fraction cubes and fraction tiles typically come prelabeled and thus predefine the whole. Again, this may limit students' mathematical flexibility, creativity, and opportunity to learn about different wholes and parts.

Once students have the comparison of part-to-whole concepts using the whole as the largest complete rod, tower, or tile, it is time to move beyond unit fractions by discussing the size of each part in relationship to the whole.

Introducing Fractions That Represent Parts Larger Than the Whole

While the focus has been on a larger rod, tower, or line being larger than the part, teachers can also begin to focus more on students' understanding of how parts can be bigger than a whole. For example, a teacher may ask a student to put together rods, towers, or lines that are larger than one and ask how much larger. For example, if my number line is 1, a teacher may ask students, "How many $\frac{1}{3}$ pieces need to come together to be larger than 1?" [Answer: 4.] Or, "How many $\frac{1}{4}$ tiles need to come together to be larger than 1?" [Answer: 5.] This simple change in words allows students to begin to explore that parts can be larger than the whole (i.e., $\frac{5}{4}$ is larger than 1). This serves to introduce not only mixed fractions but also more complex understanding of the measurement/length models and of fractions with one continuous whole/object/length. Students who can fluidly move between part–whole relationships are ready to explore discrete models for fractions that involve the whole as a collection of objects.

Discrete Models

Discrete models are part of the area of mathematics that explores finite countable sets, including cardinality, natural numbers, and integers—topics that students might already be familiar with from their PreK–2 experiences. The discrete models for teaching fractions will always involve the counting principles, natural numbers, operations on whole numbers, and the students' ability to use their number sense to solve these problems. For example, students at very young ages provide correct reasoning for why 3 is half of 6: They may say that it takes two 3s to make 6, using the "fair share" division strategy, where students break 6 into two equal groups of 3 objects each. Teachers may extend these questions into Grade 2–3 fraction activities by asking students to use counters to identify how large groups can combine into equal smaller groups.

For example, a teacher may ask a student to identify the number of counters (Figure 3.6) that comprise a third (or fourth) of the total set of counters, such as: "If 12 counters make a whole set, what fraction of the whole set is 9 counters?" [Answer: $\frac{3}{4}$.] Or: "If 12 counters are three fourths of a whole set, how many counters are in a whole set?" [Answer: 16.] These types of questions require students to shift their reasoning from tactual and visual representations of a continuous whole to using mental math strategies and counting, as well as multiplicative reasoning, to arrange the counters correctly by representing both the numerator and the denominator. (Note: While counters typically used in the classroom are similar to those in Figure 3.6, any manipulatives work in this scenario, as long as they are the same. For example, students can use 12 pennies, 12 blocks, or 12 plastic counting bears. Practice should emphasize the mathematics, not the type of manipulative.)

Figure 3.6 Counters

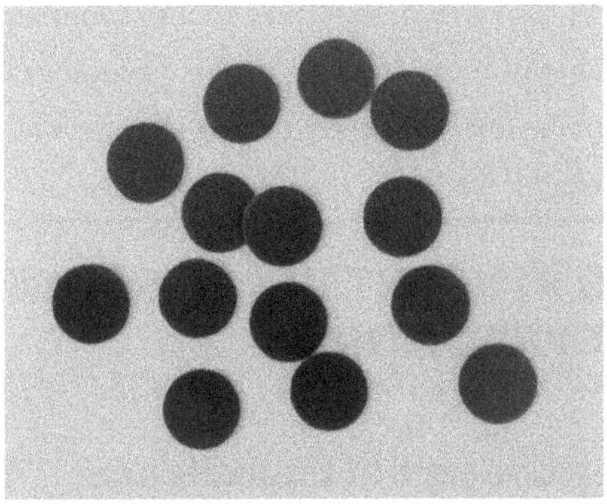

Counters are a basic and versatile manipulative.

Once again, after students show great progress and conceptual understanding of discrete models, teachers can continue posing extension questions that involve parts larger than the whole; this will allow students to further explore improper fractions using another hands-on tool. For example, teachers can ask the following: "If 15 counters make a whole set, how many counters make five thirds of a whole set?" [Answer: 25.] And: "If 12 counters make up four thirds of a whole set, how many counters are in a whole set?" [Answer: 9.]

Developing students' understanding of fractions takes a considerable amount of time and patience, and it is unquestionably an accumulative learning experience that helps students build upon meaningful, conceptual understanding via carefully structured activities, questions, and opportunities that involve different types of fractions, representations, and models. Once students have developed a strong conceptual understanding of fractions, including unit and nonunit fractions and improper fractions, they are ready to explore *equivalent* fractions.

Sidebar 3.3 Accessible Number Lines

To expand on the manipulatives presented in the chapter, teachers can also incorporate the Desktop Stick-On Number Lines, available from APH at aph.org/product/desktop-stick-on-number-lines-large-print-braille-5-pack. This product includes large print and braille to visually and tactually demonstrate number concepts, helping students make explicit connections between whole numbers and fractions (as well as mixed numbers and decimals) through concrete representations of placements of fractions on the number line. APH also produces Consumable Number Lines (aph.org/product/consumable-number-lines) as well as a Number Line Device (aph.org/product/number-line-device) that allows braille readers to more easily use accessible number lines.

DEVELOPING STUDENTS' UNDERSTANDING OF EQUIVALENT FRACTIONS

Introducing the Concept of Equivalence

To introduce *equivalence*, teachers must begin with the explorations that allow students to make sense of why certain pieces or parts of the same whole are equivalent. For example, if a person divides a pie or circle into four equal parts, why are two fourths of the circle equal to one half? Teachers should have students informally describe this equivalence relation. It is also important to continue to provide students with opportunities to use hands-on activities with concrete objects, manipulatives, and cutouts; this will allow students the opportunity to generate informal dialogue or "proofs" for the congruence of equivalent parts.

Research shows that most teachers do not spend enough time on developing the necessary reasoning and skills related to and required to conceptually understand equivalence of fractions; instead, they hastily jump into symbolism, abstraction, procedures, and rules (Behr, Wachsmuth, Post, & Lesh, 1984; Pedersen & Bjerre, 2021; Wong & Evans, 2007). Therefore, it is crucial that teachers conduct informal proofs with both dialogue and hands-on manipulatives to deepen students' understanding of both fractions and geometry, helping them to develop the logical reasoning required to truly understand equivalence relation.

Types of Equivalence

Mathematically, equivalence involves three types of relations: reflexive, symmetric, and transitive.

Reflexive Relations

The first equivalence relation type, *reflective,* is quite easy for students to understand because it involves an element (A) that can be a number or a fraction that is equal to itself ($A = A$). Using objects, students can easily explain and demonstrate why 5 = 5 or why two sets have the same number of objects.

Symmetric Relations

The second equivalence relation type, *symmetric,* is a bit more involved than the reflexive relation and includes two elements (A and B) that belong to the same equivalence class (that is, the same whole) such that $A = B$ and $B = A$. For example, if a teacher provides a circle manipulative cut into four equal pieces, students can tactually discover that two pieces of the circle equal the same size as when the remaining pieces come together—essentially, 2 half circles equal 2 half circles. Combining tactual pieces allows students to provide a visual or tactual explanation, which teachers should encourage students to verbalize. Similarly, two right triangles of equal size can come together to create one rectangle. These shapes can illustrate how more than one shape can come together to create a whole when all shapes are equal. To explicitly relate these observations to *fraction equivalence,* teachers should focus students' attention and language on the fact that that the two halves are "parts" of the same circle or shape (that is, the same equivalence class). In these examples, it is essential to include student reasoning about why a half of a circle and a half of a square (or half of a larger or smaller circle) may not be equivalent or equal. This detail is mathematically vital yet often overlooked when teaching students about *fraction equivalence.*

Transitive Relations

The third equivalence relation type, *transitive,* uses what students may already have learned during their explorations of reflective and symmetric equivalence. For this property, students will compare three elements (A, B, and C) to show that if $A = B$ and $B = C$, then $A = C$. Teachers will need to rely upon tactual manipulatives that are not equal in size. It is important to start with equal pieces coming together as a whole (2 half circles becoming 1 whole circle) before introducing that different parts can be equal (1 half circle and 2 quarter pieces of a circle making a whole). During these explorations, students should attempt to show/explain why two fourths is congruent or equivalent to a half and continue these discussions into why three sixths is congruent or equivalent to a half. Again, students can explore this by using different manipulatives of shapes and pieces, moving beyond those labeled with numeric values to ensure true mathematical understanding. Students need to also be able to identify other fractions or parts that are equivalent or congruent to a half of the same whole. The teaching and learning must focus on geometric reasoning, pictures or manipulatives, and hands-on informal proofs of equivalence, using the correct language and terminology.

In demonstrating transitive relation with a circle, a teacher needs to have a tactile piece that represents half of the circle, and then two additional pieces that are one quarter of the circle—essentially, another half piece of the circle cut in half. The teacher should then ask the students how to make a circle given the three pieces (the half, and two quarter pieces). Teachers should ask students to show and explain why the half circle is equivalent to the two quarter pieces of the circle. Additionally, teachers should encourage students to use tactual manipulatives to show and explain why differently shaped parts are congruent/equal to the same larger shape, essentially demonstrating the mathematical relationship of the *transitive relation*, showing an emerging understanding how fractions can exhibit equivalence.

Introducing the Written Fraction

Once students have shown a strong understanding of equivalence using shapes, teachers can begin to introduce *fraction notation*. With this, students label their work (defining the parts of a circle as $\frac{1}{2}$ and $\frac{1}{4}$, for example), become more proficient in symbolic representations of fractions as algebraic equations and expressions denoting fraction equivalence, and use manipulatives to understand and explain this understanding. Such manipulatives include number lines, fraction tables, and fraction tiles with fractions written in large print and UEB Math/Science or Nemeth code directly onto the manipulative. Students should continue to observe fraction equivalence through math *play*, using mathematical tools and representations to support their learning.

Once students have mastered the meaning of equivalent fractions, it is important they begin writing them on paper, using a print device, or with a braillewriter to show understanding of proper notation and equivalence. (For instance, students should be able to write $\frac{2}{4} = \frac{1}{2}$ or $\frac{3}{6} = \frac{1}{2}$). When teaching fraction notation in Nemeth code or UEB Math/Science, introduce games that enable students to spot errors and further understand the written notation. (Project Inspire has great games that focus on writing and finding errors that students often make when writing fractions. Course 3 in the series, found at pathstoliteracy.org/Nemeth-Code-Symbols-Used-Grades-2-5-Course-3, will help teachers refresh their notation-writing skills.)

DEVELOPING STUDENTS' UNDERSTANDING OF ADDITION/SUBTRACTION OF FRACTIONS

Teaching addition and subtraction of fractions begins in Grade 4, when students receive an official introduction to the definition of a fraction as a *ratio* between the *numerator* and *denominator*. Initially, students add and subtract fractions with "like" denominators, as this is the fundamental concept behind addition and subtraction of fractions; if students do not grasp this concept, misconceptions can carry over into middle and high school mathematics. Hence, teachers must continue building upon students' geometric reasoning for understanding addition and subtraction of fractions, asking students to use manipulatives and models to explain or show how the *size* of each piece and the *quantity* of pieces they added or subtracted affect the final answer. Again, teachers should not rush this learning process or replace it with symbolism, procedures, rules, or algebraic notations before students are ready. It does not help students learn, nor does it help them make much progress.

Teaching Using Manipulatives

When teaching addition and subtraction of fractions, teachers can use fraction circles or fraction tiles and ask students to manually move the pieces together or apart with "like" denominators to explain and make sense of the answer. For example, students can explain why $\frac{1}{3} + \frac{1}{3}$ is not $\frac{2}{6}$. This misconception of adding numerators and denominators is the most common among students (Ashlock, 2006), making it a great example to discuss. Teachers must spend considerable instructional time using counterexamples to show *why* adding numerators and denominators neither works nor makes sense when adding fractions. One of the most powerful strategies is to use fraction manipulatives (fraction circles or fraction tiles) and make sure these manipulatives come with fractions of consecutive denominators.

Figure 3.7 Fraction Circles Demonstration

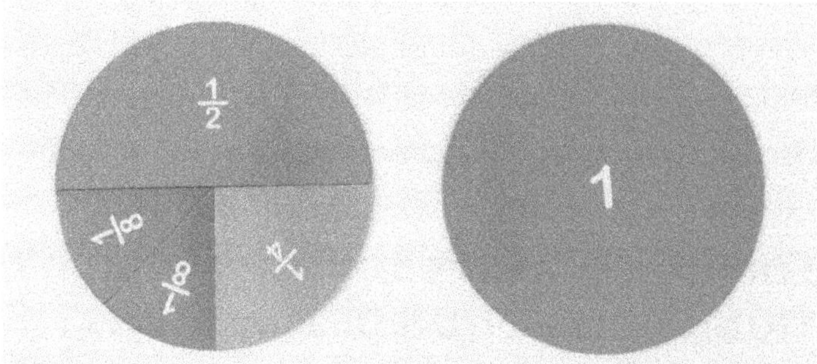

Most fraction manipulatives, like these fraction circles, feature consecutive denominators. They can be used to teach the relationships between whole numbers and fractions.

Fraction tiles are long rectangles with markings that show fractional parts. For example, the most basic strip showing a fraction may display labels reading $\frac{1}{2}$ and $\frac{1}{2}$ opposite an equal point of bisection; that same strip is the same length as the whole. Teachers can label the fraction tiles (including with tactile indicators and braille so students can discern fractional parts) and divide them into many fractional parts, but the parts still add up to the whole.

Fraction circles, shown above, are similar to fraction lines. All the circles are the same size, but each circle features different fractional components, such as $\frac{1}{2}$ with two equal parts to $\frac{1}{12}$ on another circle with 12 equal parts.

Sidebar 3.4 Creating Fraction Manipulatives

Teachers can easily convert fraction manipulatives into tactual manipulatives using cardstock or foam and a braillewriter or marker to label each part. In addition, APH offers an accessible app, Practice2Master Fractions (aph.org/product/practice2master-fractions) that students can use on their own devices to examine representations of fractions and aid in calculations. Magnetic fraction circles and fraction bars are also available, but teachers will need to label each fraction in either UEB Math/Science or Nemeth code.

When choosing fraction manipulatives, teachers should include manipulatives with consecutive denominators to help students mathematically observe that the denominator *always* represents the "size" of the parts and the numerator represents the "quantity" of the (equal-sized) parts. To help strengthen this understanding, teachers should repeatedly ask students to provide explanations when answering questions, such as:

- Why do we need exactly seven $\frac{1}{7}$ pieces to make 1 whole circle? [Answer: It varies but could include that each piece is $\frac{1}{7}$ of the total circle, or each of the 7 pieces represents $\frac{1}{7}$ of the total circle.]
- How many ninths pieces do we need to make 1 whole circle? [Answer: 9.]
- If we want to make 3 whole circles, how many ninths do we need? [Answer: 27.]

Teachers can continue this line of questioning to encourage deeper understanding and allow students to demonstrate the ability to explain parts to whole and equivalency. Questions should increase in complexity, such as: "How many whole circles can we make with 13 fifths? Would there be any left over?" [Answer: 2 circles, with 3 pieces left over.] Teachers should provide each student with physical manipulatives (fraction circles or fraction tiles) to help them answer these questions.

Overcoming Common Misconceptions Using Manipulatives

After students have demonstrated proficiency in answering these questions, they might be ready for more follow-up questions to initiate their thinking about adding and subtracting fractions with "unlike" denominators: "How can we add $\frac{1}{3}$ and $\frac{1}{4}$? What's your strategy?" Allow students to discuss different strategies and ways to solve the problem.

However, teachers need to be ready to confront common misconceptions that arise when adding fractions with different denominators, such as $\frac{1}{3} + \frac{1}{4} = \frac{2}{7}$. To do so, teachers can explicitly show students, using manipulatives, that "two sevenths" placed on top of $\frac{1}{3} + \frac{1}{4}$ is not equal. This helps students recognize that the "surface" or "area" occupied by two sevenths does not equal the "surface" or "area" of $\frac{1}{3} + \frac{1}{4}$ combined. Students can also explain why the "size" of sevenths is much smaller than the "size" of thirds and fourths added together. Encouraging the use of manipulatives to correct misconceptions early is vital and allows students to recognize and remedy their own misconceptions.

Introducing Adding and Subtracting Fractions With Unlike Denominators Using Manipulatives

After students have shown a strong conceptual understanding of fraction equivalence, addition and subtraction of fractions, and why $\frac{1}{3} + \frac{1}{4}$ is not $\frac{2}{7}$, they are ready to grapple with the questions: "How can we add $\frac{1}{3}$ and $\frac{1}{4}$? What's your strategy?" If students are unsure, teachers may offer a reflection on their previous experiences of easily being able to add fractions with "like" denominators, asking students to find another equivalent fractional piece that can represent both thirds and fourths and serve as the "like" denominator for both. Again, during these explorations, teachers should continue building upon students' geometric reasoning and tactile experiences without teaching rules, symbolism, or procedures. By using tactile manipulatives such as fraction circles and fraction tiles, students begin to make sense of the importance of the fractions' size and understand why finding another equivalent fraction for both is vital, as is a shared or common or "like" denominator for adding these fractions with "unlike" denominators.

Elementary-grade students often recognize the need for equivalent fractions (or a common denominator) in order to add two fractions with "unlike" denominators, and the largest struggle typically involves the process of finding such equivalent fractions. Hence, teachers should provide sufficient time for students to grapple with this process, using physical models to add and subtract fractions with "unlike" denominators. In fact, research shows that most Grade 4–5 students, via this process and with the help of concrete or virtual manipulatives, begin to notice mathematical patterns and develop their own techniques for adding and subtracting fractions (Burns & Hamm, 2011; Gaetano, 2014; Reimer & Moyer, 2005).

Moving Beyond the Manipulatives: Using Mathematical Strategies for Adding and Subtracting Fractions With Unlike Denominators

After extensive explorations of fraction addition/subtraction using manipulatives, mathematical strategies can be introduced.

Method 1: Any Common Denominator

Students who use Method 1 mainly focus on finding any common denominator by using an "easy" multiplication fact that both denominators share. For example, in a problem examining $\frac{2}{3} + \frac{1}{6}$, a student may notice that they can split both the denominators into twelfths. A student may count by 3s or 6s to determine that each number has a common denominator of 12. While the student did not consider the least common multiple, the student nonetheless did find a common number between the two. From there, they can multiply each fraction to arrive at the common denominator: $\frac{2}{3}$ becomes $\frac{8}{12}$, and $\frac{1}{6}$ becomes $\frac{2}{12}$. In adding the two numerators together, as depicted in Figure 3.8, the student ultimately gets the answer of $\frac{10}{12}$.

Figure 3.8 The Any Common Denominator Method

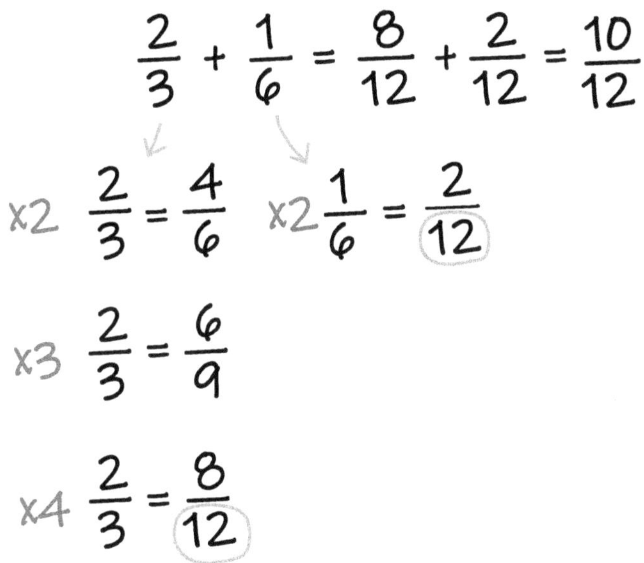

Here, the student has found the common denominator of 12 and multiplied the fractions accordingly.

Method 2: Multiplying Denominators

Students who use Method 2 do not consider the least common multiple; their method involves finding the common multiple by multiplying the denominators. In the example problem, students attempt to add $\frac{2}{3}+\frac{1}{6}$ and find the common denominator by multiplying both denominators together to get 18. Then, the students find the numerator values by assessing how to get the denominators to 18. Thus, $\frac{2}{3}$ becomes $\frac{12}{18}$ and $\frac{1}{6}$ becomes $\frac{3}{18}$. When added together, the student receives the answer $\frac{15}{18}$. Students developed this method on their own based on mathematical observations from several previous examples, including fractions with coprime denominators (meaning they have no common factors other than 1).

Figure 3.9 The Multiplying Denominators Method

$$\frac{2}{3}+\frac{1}{6}=\frac{12}{18}+\frac{3}{18}=\frac{15}{18}$$

$$\text{common} = 3 \times 6 = 18$$

$$\frac{2\times 6}{3\times 6}=\frac{12}{18}$$

$$\frac{1\times 3}{6\times 3}=\frac{3}{18}$$

By multiplying denominators, the student is able to quickly find a common denominator that works for each fraction.

Method 3: Finding the Least Common Multiple

Students using Method 3 are able to recognize, through multiple experiences, that they can obtain the common denominator by finding the least common multiple, yielding the answer in its simplest fraction form. When using Method 3, the students notice that the least common multiple between the denominators of 3 and 6 is 6. Therefore, $\frac{2}{3}$ becomes $\frac{4}{6}$, because the students change both the numerator and denominator, so the fraction has a common denominator of 6. The second number, $\frac{1}{6}$, does not need to change. This student solves the same problem posed by adding $\frac{4}{6} + \frac{1}{6}$ to equal $\frac{5}{6}$.

Figure 3.10 The Least Common Multiple Method

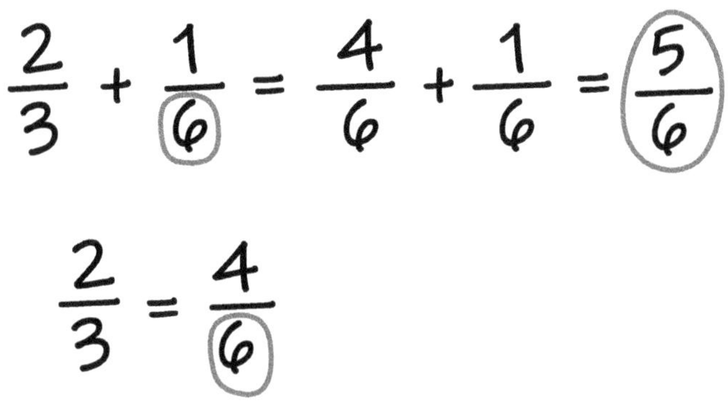

A student using the least common multiple method will arrive at the simplest fraction in the solution.

Developing Understanding of Adding Fractions

Most students concern themselves with the goal of the fraction-adding activity: finding any common denominator that will allow them to add two fractions. This is why we strongly discourage teachers from prematurely introducing the rules and procedures for adding and subtracting fractions, as those rules involve the least common denominators—which are neither intuitive nor important to students during their initial explorations. Teachers should not focus on the language and the terms "common" or "least common" multiples at this stage of mathematical exploration because the goal is for students to develop understanding and their own strategies for adding two fractions with unlike denominators.

Rather, it is best to continue building on students' own developed strategies and understanding and to keep challenging them to consider deeper questions that build upon their thinking, such as: "When adding two fractions with unlike denominators, do you think it is always going to be possible to find a common denominator for both? Why do you think so?" Allow for differing answers among the students in order to determine overall concept understanding and to begin to introduce the concept of the least common denominator. Questions such as these help students to discuss, explore, and develop a deeper understanding of concepts involving common multiples without introducing the vocabulary. The goal is for students to test a few fractions, recognize that the numerators play no part in finding the common denominator, and understand that any two or more whole numbers (denominators), including prime numbers, will always share a common multiple.

The next follow-up question series for teachers to ask could be: "When adding $\frac{2}{3} + \frac{1}{6}$, our class used different methods and got three different answers: $\frac{5}{6}$ and $\frac{10}{12}$ and $\frac{15}{18}$. Which answer is correct? How do you know?" These questions help students build upon their understanding of fraction equivalence and examine their answers closely. These types of questions and discussions should continue in the classroom to help students add and subtract fractions, develop a deeper understanding of the least common multiple (and/or the least common denominator), and practice writing their answers (regardless of solution method) in simplified fraction form.

DEVELOPING STUDENTS' UNDERSTANDING OF MULTIPLICATION/DIVISION OF FRACTIONS

Multiplication and division of fractions predominantly rest on students' geometric reasoning about area, as well as their number sense related to multiplication and division. This is because fraction multiplication and division require students to reason portions of "parts" as they relate to the whole (area).

Teaching Multiplication of Fractions

Method 1: Multiplication Using Part-to-Whole or Group-to-Quantity

When teaching *multiplication of fractions*, it is important to connect students' learning to the meaning of multiplication. Teachers should avoid using terms such as "factor" and "product" when teaching fraction multiplication. The reasoning behind this is due to the nominal focus on the operation (multiplication) and not the *part-to-whole* or *group-to-quantity* relationships required to conceptually understand multiplication and division of fractions. Hence, the first step in teaching fraction multiplication (and division) is to focus students' learning on two easy cases using whole numbers. For example, students can begin by exploring and then explaining why 4 groups of $\frac{1}{3}$ (or $4 \times \frac{1}{3}$) would require repeated addition of $\frac{1}{3}$ four times. These types of examples help students make sense of fraction multiplication as "groups" of $\frac{1}{3}$.

Method 2: Using Geometric Reasoning

Students will find the second method of fraction multiplication more mathematically involved, as it requires them to use geometric reasoning to partition a whole (number) or a quantity into fractional groups or parts, such as $\frac{1}{3}$ groups of 6 or "one third of six" ($\frac{1}{3} \times 6$). Students can use any manipulatives (e.g., counters, number line, fraction tiles, fraction circles) to make sense of both the structure and the solution. For example, using fractions and parts by taking a third from each whole, students will arrive at six groups of $\frac{1}{3}$s, or $\frac{6}{3}$. On the other hand, students could use 6 counters (no fractions) and split the 6 into 3 groups (or thirds), which will result in 2 (whole) counters, no fractions. Both answers, $\frac{6}{3}$ and 2, are correct due to equivalence. This merits an explicit discussion during this lesson—teachers can open this lesson with a story problem about splitting 6 cookies among 3 friends and ask students to explain: "How is '$\frac{1}{3}$ of 6' or '$\frac{1}{3} \times 6$' related to this story problem?" (Teachers could even provide students with six cookies to help them tactually demonstrate their solutions, discover both solutions, and discuss the two equivalent solutions.)

Again, teachers must not rush into introducing rules or procedures for fraction multiplication, particularly because research shows that telling students to "multiply numerators and denominators across" when multiplying fractions lacks meaning and effectiveness and often yields incorrect work, such as: $4 \times \frac{1}{3} = \frac{4}{12}$ or $\frac{1}{3} \times 6 = \frac{1}{18}$ (Ashlock, 2006).

Teaching Multiplication of Fractions by Whole Numbers

Once students have demonstrated full proficiency in multiplying unit fractions by whole numbers, teachers may build upon this understanding to challenge them further: "Using manipulatives, show half of $\frac{1}{3}$ or $\frac{1}{2} \times \frac{1}{3}$. What is your strategy?" These questions help students make sense of fraction multiplication and demonstrate their understanding of the following: (1) partitioning a third into two equal parts (halves); (2) correctly labeling each new part in terms of the whole as a result of partitioning; and (3) identifying the region that represents "half of a third" or $\frac{1}{2} \times \frac{1}{3}$ and the numbers in the numerator and denominator. Teachers need to keep in mind when teaching fraction multiplication that students will need ample opportunities to keep working with these types of problems and manipulatives to gradually develop a deep, meaningful, and conceptual understanding of *fraction multiplication*.

Ashlock (2006) found that adults, including teachers, think about the multiplication of fractions fundamentally differently than do students, and thus the way many teachers teach fraction multiplication can be counterintuitive to the way students learn and think about partitioning fractions. Students often learn that they should multiply across numerators and denominators, but research (Ashlock, 2006) shows that students do not notice multiplication as an operation between numerators and denominators. Once students demonstrate a strong conceptual understanding of fraction multiplication, they begin to notice specific patterns and create their own "rules," making connections in their answers regarding numerators and denominators that they can multiply.

DIVISION OF FRACTIONS

Similar to multiplication, *division of fractions* involves students' geometric reasoning and understanding of the division structure. Hence, teachers should follow a similar approach to teaching division of fractions as fraction multiplication, focusing the initial instruction on two easier cases: *dividing a (unit) fraction by a whole number* and *dividing a whole number by a (unit) fraction*. Both cases will continue to involve what students already understand from their previous learning experiences with whole numbers and fraction multiplication.

Starting With Dividing a Unit Fraction by a Whole Number

At first, the teacher will begin by introducing division as a problem that involves *dividing a (unit) fraction by a whole number* and finding the size or quantity of each group. When beginning this method, teachers will want to ensure that the fraction stays the same as the divisor changes. This is deliberate, as the goal is to help students make sense of the overall division structure as it relates to the *fraction division* concept—*not* to engage students in any computations yet. It is critical that students have mathematics manipulatives (such as fraction tiles) to aid their learning, problem-solving process, and solutions. Students should have access to fraction manipulatives that include consecutive denominators for mathematical flexibility.

This particular method of fraction division builds directly upon students' understanding of fraction multiplication, as it requires them to partition the given "parts" into equal groups. Specifically, teachers should ask students to explain:

- How can we divide a third into 2 equal parts? Into 3 equal parts? Into 4 equal parts?
- Use the fraction tiles to find another fraction piece that exactly represents the answer for each case. How does your answer (fraction piece you found) relate to the fraction and the divisor?
- What patterns do you notice?

Exploring several examples can help students recognize that when they divide a (unit) fraction by a whole number and partition a given "part" into equal groups, they should write the answer in terms of the whole and not the given part.

Using Manipulatives to Divide Fractions

In Figure 3.11, students should receive a fraction tile ($\frac{1}{3}$) and try to find another fraction tile that represents exactly one half of $\frac{1}{3}$. For students with visual impairments, they will need access to accessible fraction tiles. Students will be able to find the $\frac{1}{6}$ tile and notice that two $\frac{1}{6}$ tiles equal $\frac{1}{3}$, and teachers should ask them to define it as a fraction, without giving them the answer. These types of problems can build on students' understanding of division (such as $12 \div 2 = 6$), as well as division story problems about "evenly splitting 12 cookies into 2 groups," where students define how many in each group. In this case, students engage in the same process using fractions: taking $\frac{1}{3}$, evenly splitting it into 2 groups, and defining the number of items in each group.

Figure 3.11 Dividing a Unit Fraction by a Whole Number

$$\frac{1}{3} \div 2 = ?$$

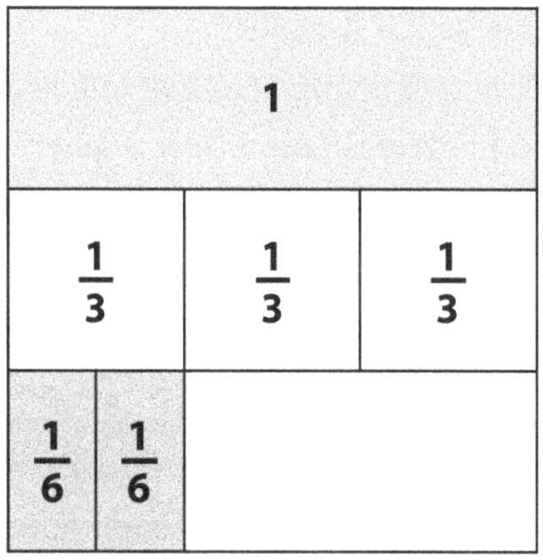

Completing this problem with manipulatives allows students to understand the sometimes-counterintuitive division of fractions.

Once students show proficiency and comfort in correctly answering these types of mathematical questions using manipulatives, teachers can pose extension questions that help students gradually move away from "hands-on" explorations to more "minds-on" connections, engaging them in considering these questions:

- In each of these problems, how is the answer related to the numbers in the original problem? $\frac{1}{3} \div 3 = \frac{1}{9}$, $\frac{1}{3} \div 4 = \frac{1}{12}$, $\frac{1}{3} \div 5 = \frac{1}{15}$
- What do you notice?

As mentioned earlier, given that students learn division of fractions in Grade 5 after fraction multiplication, most fifth graders can begin to make mathematical connections, such as that $\frac{1}{3} \div 2 = \frac{1}{6}$ mirrors the multiplication problem $\frac{1}{3} \times \frac{1}{2} = \frac{1}{6}$ and, similarly, the problem $\frac{1}{3} \div 3 = \frac{1}{9}$ is directly related to the multiplication problem $\frac{1}{3} \times \frac{1}{3} = \frac{1}{9}$. At this stage of mathematical development, teachers should ask students what they notice and for any strategies and conjectures they may have; students can then share their discoveries with the class, writing their ideas onto a poster to record and revisit later.

Using Manipulatives for Division of a Whole Number

The second type of division of fractions involves *dividing a whole number by a (unit) fraction*—in other words, finding the number of groups (A), such as: $1 \div \frac{1}{3}$ and $2 \div \frac{1}{3}$ and $3 \div \frac{1}{3}$. Please note that in this case, the divisor remains the same, as the goal is to help students understand the concept of what it means to divide by a fraction rather than engage in computations. Similarly, for this case, it is vital that students use mathematics manipulatives (e.g., fraction tiles) to support their explorations, thinking, and problem solving. It is also important for the teachers to continue resisting the desire to show or tell students any rules and symbolic manipulations for division of fractions.

Most fifth graders show very strong readiness for this particular case of fraction division and often do very well in solving these problems with the use of manipulatives, and without much help from teachers. Figure 3.12 is an example of the intuitiveness of the mathematical process behind this type of fraction division.

Figure 3.12 Dividing a Whole Number by a Unit Fraction

Problem A: $1 \div \frac{1}{3} = 3$

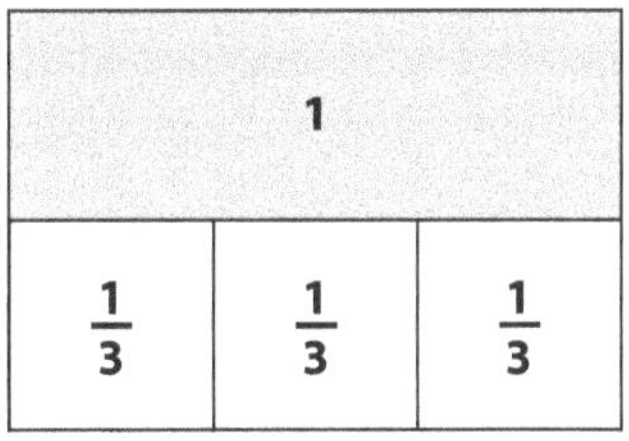

Problem B: $2 \div \frac{1}{3} = 6$

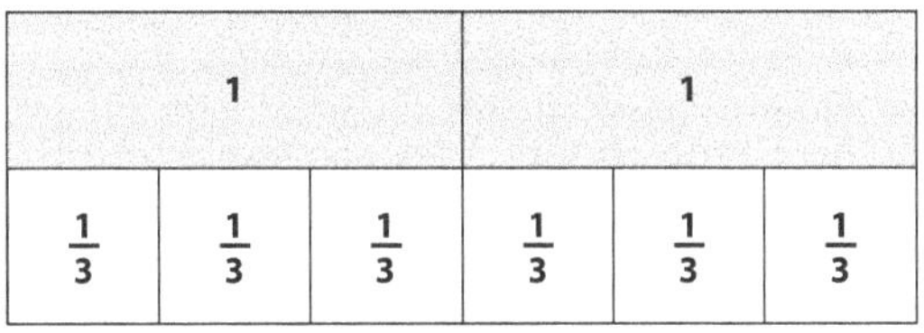

Dividing whole numbers using manipulatives provides an easy way for students to develop understanding, and beginning by keeping the fraction the same will help students see the foundational concept before engaging in computation.

For Problem A in Figure 3.12, most fifth graders find it easy to explain why there are three of the fraction $\frac{1}{3}$ in a whole. Students place three $\frac{1}{3}$ tiles under the 1 whole tile to represent the division problem. Hence, they easily can provide an explanation for any of the problems similar to Problem A involving any unit fraction as a divisor. Starting the lesson by posing Problem A, which asks students to divide 1 into fractions, establishes the mathematical foundation upon which students are ready and expected to build the rest of this case: Problem B and repeated addition of groups of $\frac{1}{3}$. Again, to help students focus on the concept more closely and make sense of Problem B, teachers need to follow up with specific questions:

- Use the fraction tiles to show how many thirds go into 1 whole ($1 \div \frac{1}{3} = 3$)?
- How about 2 wholes ($2 \div \frac{1}{3} = 6$)? Three wholes ($3 \div \frac{1}{3} = 9$)?
- How many thirds do you think would go into 6 wholes ($6 \div \frac{1}{3} = 18$)?
- Ten wholes ($10 \div \frac{1}{3} = 30$)?
- How do you know? What is your strategy?

When written symbolically, the problem often can confuse students, so posing these questions helps students recognize what the question investigates: "How many groups of $\frac{1}{3}$ can fit into (or go into) the total number of wholes?" Students can draw upon their earlier experiences with whole number division (such as $12 \div 2 = 6$ or $12 \div 4 = 3$), story problems involving "splitting 12 cookies into equal groups with 2 or 4 cookies in each group," determining the number of needed groups. These questions also offer students the flexibility to build wholes with thirds and use repeated subtraction or addition to count how many groups of thirds they need to make the total number of wholes.

Furthermore, these problems are very accessible to fifth graders and can straightforwardly lead them to notice the "invert and multiply" rule in a meaningful and hands-on way. Many students can make conjectures, such as: "Because there are 3 thirds in every whole, there will be 6 thirds in 2 wholes, and 9 thirds in 3 wholes, and (6×3) 18 thirds in 6 wholes.

To help students make further connections, teachers can add extensions of fraction division using nonunit fractions and spend some time discussing and reviewing the meanings of both *unit fractions* and the *reciprocals of all fractions*.

This can help students make more accurate conjectures related to the "rules" they notice, extending their observations for division of *all* fractions beyond unit fractions. For example, teachers can pose a problem: "Use fraction tiles to find how many groups of "two thirds" will go into 4 wholes." (Please note that this question for fraction division deliberately does not involve any remainders and allows students to easily build upon their explorations by using manipulatives.) It may also help to provide students with adhesive tape to stick together the two orange pieces (one thirds) to define *one group*. Using their previous work (Figure 3.12, Problem B), most fifth graders can recognize that 3 groups of two thirds make 2 wholes ($3 \times \frac{2}{3} = 2$). Therefore, double that amount will make 4 wholes (i.e., 6 groups of two thirds will make 4 wholes) and they can continue the pattern to make four wholes using two thirds. Teachers can write the following on the board, posing additional questions, such as:

- What do you notice for all the division problems $2 \div \frac{2}{3} = 3$ and $4 \div \frac{2}{3} = 6$?
- Do they follow our class rule from the previous fraction division examples?
- What is a reciprocal of $\frac{2}{3}$?
- Does our class rule still work if we multiply by the reciprocals? Is $2 \div \frac{2}{3} = 2 \times \frac{3}{2}$? $4 \div \frac{2}{3} = 4 \times \frac{3}{2}$?
- What do you notice?
- What is your rule or strategy for division of fractions?

SUPPORTING STUDENTS' ABILITY TO ANALYZE MATHEMATICAL RULES

As students and teachers continue to engage in meaningful learning and teaching activities, students will make stronger connections between the two (easier) division-of-fractions cases and extend their observations to developing a concept for division of *all* fractions. To facilitate this, teachers provide students opportunities to use manipulatives to support their thinking and problem solving, find solutions to these types of questions, and check their solutions using their own "rules" developed and agreed upon as a class. Teachers also should prompt students to keep checking *if*, *why*, and *how* their own rules work for *all* fraction division cases. This process will require patience in allowing students the needed time to keep exploring and testing their conjectures, rules, and hypotheses against their mathematical observations.

Further, teachers must offer students the opportunity to be able to *observe* and *generalize* their rules and conjectures about *division-of-fractions* cases and refine their rules to make sure they work for *any* fraction division case. After students are ready to reach an agreement on the rule that they all confirmed works for various fraction division cases, teachers should provide fifth graders with calculators and not be afraid to pose more complex fraction division problems to test this rule. Teachers should keep in mind that when teaching fractions, computation plays a small part in the process; comprehension and conception complete students' understanding of fractions.

EXTENDING STUDENTS' UNDERSTANDING OF FRACTIONS TO DECIMALS

Students' initial introduction to decimals as "money" comes around Grade 2, when they learn about coin values, making change, and currency-based quantities that involve whole dollars and fractions of the dollar (cents).

These experiences often help students make sense of decimals in Grade 4–5, when they begin to explore decimals as base-10 numbers. In fact, these experiences truly help students add and subtract decimals by applying what they already learned in vertical addition and subtraction of whole numbers by preserving the position (aligning) the decimal points. By Grade 3, most students understand why 2.34 is less than 2.43, because they connect their reasoning to money. Also, students are able to apply their previous knowledge about whole numbers to develop an understanding of rounding decimals.

Misconceptions in the Understanding of Decimals

The extensions of fractions into decimals discussed in this chapter specifically relate to the three research-based misconceptions (Irwin, 2001; Resnick et al., 1989) rooted in students' prior mathematical experiences with *whole numbers*, the role of *zero*, and the misapplication of *language*. Specifically, the *whole-number* misconception involves students' belief that 0.6 is less than 0.26 because 6 is less than 26 (Ni & Zhou, 2005). The *role of zero* misconception relates to students' assumption that zero means "nothing" or has "no value" or "no quantity," and so 0.7 and 0.07 and 0.007 must be the same. The role of zero misconception is also problematic for the cases when zero falls at the end of the decimal, as some students believe that 0.32 is less than 0.320 because 32 pennies is less than 320 pennies (Durkin & Rittle-Johnson, 2012; Irwin, 2001). The misapplication of language is shown as students assume that 0.32 is greater than 0.320 because hundredths are greater than thousandths (Irwin, 2001; Resnick et al., 1989).

Using Manipulatives to Overcome Misconceptions

To address these misconceptions, teachers should engage students in conceptual understanding of decimals and not focus on rules, tricks, or procedures. Rather, teachers should offer students opportunities to engage in conversations about place value, specifically the value of each decimal place separately. Furthermore, teachers should extend and expand students' understanding of fraction equivalence to build *decimal equivalence* (0.32 = 3 tenths + 2 hundredths = 30 hundredths + 2 hundredths; also $\mathbf{0.32} = \frac{3}{10} + \frac{2}{100} = \frac{30}{100} + \frac{2}{100} = \frac{32}{100}$). Doing so teaches students how to write decimals in *expanded form* and offers conceptual ways for them to compare decimals. For example, when comparing 0.30 and 0.3, students can easily observe that 0.30 = 3 tenths + 0 hundredths and 0.3 = 3 tenths and quickly notice that 0.30 = 0.3 has the same number of tenths and hundredths.

Several products available through APH further support overcoming misconceptions. The Flip-Over Books, available for both Nemeth code and UEB Math/Science, allow students to flip over tactile representations of fractions and decimals to determine the equivalency of fractions and decimals specifically written in both large print and UEB Math/Science (aph.org/product/flip-over-concept-book-fractions-ueb) or Nemeth code (aph.org/product/flip-over-concept-book-fractions-nemeth). Additionally, Digi-Blocks (louisekool.com/collections/digi-blocks) provide hands-on opportunities for students to determine equivalency in fractions and decimals. (Videos with ideas for specific lessons plans and using these manipulatives is available at youtube.com/user/digiblock.)

Understanding Decimals as Fractions

Understanding the *equivalence of decimals* and writing decimals in *expanded form using fractions* (example: $0.324 = \frac{3}{10} + \frac{2}{100} + \frac{4}{1000}$ and also $0.324 = \frac{324}{1000}$) is critical to students' understanding of operations on decimals, especially multiplication and division. When teaching multiplication and division of decimals, teachers often focus heavily on rules, procedures, or tricks (such as shifting the decimal point) and do not help students develop a conceptual understanding of these topics. Building upon decimal equivalence and writing decimals in expanded form using fractions is key.

Multiplying Decimals

When multiplying 0.4 × 0.08, students can rewrite the decimals as fractions and apply fraction multiplication to solve the problem: $0.4 \times 0.08 = \frac{4}{10} \times \frac{8}{100} = \frac{32}{1000} = 0.032$. In fact, math "tricks" such as shifting the decimal point often leads to errors when, due to too many math tricks to memorize, students cannot remember how many places they must shift. This is problematic particularly for problems such as 0.4 × 0.5, for which students (incorrectly) think that the decimal point shifts one decimal place value, as both 0.4 and 0.5 have one decimal place value; again, some students also assume that zero means "nothing" and the decimal shift of two places starts at the nonzero number. Both assumptions lead to incorrect answers, and a more conceptual approach to teaching decimal multiplication will help students understand and solve the problem using fraction representations: $0.4 \times 0.5 = \frac{4}{10} \times \frac{5}{10} = \frac{20}{100} = 0.20$ (or 0.2 due to decimal equivalence).

Dividing Decimals

Similarly, for division of decimals, many teachers rely primarily on teaching this concept right away by using the long division method, one of the most difficult (and frustrating) methods for young students (Ashlock, 2006). As always, teachers should build upon students' understanding of decimal equivalence and their ability to write decimals in expanded form using fractions. Specifically, when dividing 0.6 by 0.2, students can reason, "How many groups of 2 tenths fit into 6 tenths?" The answer is 3 groups, because 0.2 + 0.2 + 0.2

= 0.6. Students can apply this directly to all examples for dividing decimals of the same length.

Once students understand the division of decimals, students can either apply what they already know about the multiplication of decimals (for example: $0.56 \div 0.7 = ?$ is equivalent to $0.7 \times ? = 0.56$) or continue rewriting decimals as fractions to solve the problem ($0.56 \div 0.7 = \frac{56}{100} \div \frac{7}{10} = \frac{56 \div 7}{100 \div 10}$). These explorations build a strong conceptual foundation for students to be able to develop an understanding operations involving decimals, as well as explicitly and mathematically observe why, when dividing larger and more complex decimals using long division, they must "shift" the decimal point for specific cases.

CONCLUSION

Fractions and decimals (rational numbers) can appear daunting to the teacher and student at first, but through a combination of sound pedagogical approaches, use of manipulatives, and allowing learners to discover the rules and patterns of problem solving, students can make steady progress toward mastery. We hope that this chapter will give teachers a more pedagogically informed insight on the specific activities, math questions, and learning opportunities they can provide students. In doing so, students will have more meaningful experiences and develop a more conceptual understanding of fractions and decimals that can lead to better mathematics knowledge and proficiency in middle and high school.

We urge teachers to continue focusing their lessons on providing opportunities that will allow students to independently discover mathematical "rules" that are just as accurate, valid, and correct as traditionally taught methods. Coupled with hands-on use of manipulatives and classroom discussions to enhance their learning and experiences—making instruction fully accessible in the process—students will make better sense of numbers both conceptually and kinesthetically. The more that we can provide universal experiences in the use of manipulatives and self-discovery learning, the better mathematicians our students will become.

REFERENCES

Ashlock, R. (2006). *Error patterns in computation: Using error patterns to improve instruction*. Pearson/Merrill Prentice Hall.

Ball, D. L., Hill, H. C, & Bass, H. (2005). Who knows mathematics well enough to teach third grade, and how can we decide? *American Educator*, *5*(3), 43–46.

Behr, M. J., Wachsmuth, I., Post, T. R., & Lesh, R. (1984). Order and equivalence of rational numbers: A clinical teaching experiment. *Journal for Research in Mathematics Education*, *15*(5), 323–341. https://doi.org/10.2307/748423

Blackorby, J., & Wagner, M. (1996). Longitudinal postschool outcomes of youth with disabilities: Findings from the National Longitudinal Transition Study. *Exceptional Children*, *62*(5), 399–413.

Burns, B. A., & Hamm, E. M. (2011). A comparison of concrete and virtual manipulative use in third-and fourth-grade mathematics. *School Science and Mathematics*, *111*(6), 256–261. https://doi.org/10.1111/j.1949-8594.2011.00086.x

Cai, J., & Ni, Y. (2011). Investigating curricular effect on the teaching and learning of mathematics in a cultural context: Theoretical and methodological considerations. *International Journal of Educational Research*, *50*(2), 65–70. https://doi.org/10.1016/j.ijer.2011.06.001

Common Core State Standards Initiative (CCSSI). (2010.) *Common core state standards for mathematical practice*. National Governors Association Center for Best Practices and Council of Chief State School Officers (CCSSO).

Durkin, K., & Rittle-Johnson, B. (2012). The effectiveness of using incorrect examples to support learning about decimal magnitude. *Learning and Instruction*, *22*(3), 206–214. https://doi.org/10.1016/j.learninstruc.2011.11.001

Emerson, R. W., & Anderson, D. (2018). What mathematical images are in a typical mathematics textbook? Implications for students with visual impairments. *Journal of Visual Impairment & Blindness*, *112*(1), 20–32. https://doi.org/10.1177/0145482X1811200103

Gaetano, J. (2014). *The effectiveness of using manipulatives to teach fractions* [Master's thesis, Rowan University]. Theses and Dissertations. https://rdw.rowan.edu/etd/495

Hiebert, J., & Grouws, D. A. (2007). The effects of classroom mathematics teaching on students' learning. In F. Lester (Ed.), *Second Handbook of Research on Mathematics Teaching and Learning*, pp. 371–404.

Hunt, J. H., Taub, M., Marino, M., Duarte, A., Bentley, B., Holman, K., & Kuhlman, A. (2023). Effects of game-enhanced supplemental fraction curriculum on student engagement, fraction knowledge, and STEM interest. *Education Sciences, 13*(7), 646. https://doi.org/10.3390/educsci13070646

Irwin, K. C. (2001). Using everyday knowledge of decimals to enhance understanding. *Journal for Research in Mathematics Education, 32*(4), 27–52. https://doi.org/10.2307/749701

Mack, N. K. (2001). Building on informal knowledge through instruction in a complex content domain: Partitioning, units, and understanding multiplication of fractions. *Journal for Research in Mathematics Education, 32*(3), 267–295. https://doi.org/10.2307/749673

McLeod, T. M., & Armstrong, S. W. (1982). Learning disabilities in mathematics—skill deficits and remedial approaches at the intermediate and secondary level. *Learning Disability Quarterly, 5*(3), 305–311. https://doi.org/10.2307/1510297

Misquitta, R. (2011). A review of the literature: Fraction instruction for struggling learners in mathematics. *Learning Disabilities Research & Practice, 26*(2), 109–119. https://doi.org/10.1111/j.1540-5826.2011.00330.x

Mohyuddin, R. G., & Khalil, U. (2016). Misconceptions of students in learning mathematics at primary level. *Bulletin of Education and Research, 38*(1), 133–162.

National Mathematics Advisory Panel. (2008). *Foundations for success: The final report of the National Mathematics Advisory Panel.* U.S. Department of Education.

Ni, Y., & Zhou, Y.-D. (2005). Teaching and learning fraction and rational numbers: The origins and implications of whole number bias. *Educational Psychologist, 40*(1), 27–52. https://doi.org/10.1207/s15326985ep4001_3

Pedersen, P. L., & Bjerre, M. (2021). Two conceptions of fraction equivalence. *Educational Studies in Mathematics, 107*(1), 135–157. https://doi.org/10.1007/s10649-021-10030-7

Reimer, K., & Moyer, P. S. (2005). Third graders learn about fractions using virtual manipulatives: A classroom study. *Journal of Computers in Mathematics and Science Teaching, 24*(1), 5–25.

Resnick, L. B., Nesher, P., Leonard, F., Magone, M., Omanson, S., & Peled, I. (1989). Conceptual bases of arithmetic errors the case of decimal fractions. *Journal for Research in Mathematics Education, 20*, 8–27. https://doi.org/10.2307/749095

Rosenblum, P. & Amato, S. (2004). Preparation in and use of Nemeth braille code for mathematics by teachers of students with visual impairments. *Journal of Visual Impairment & Blindness*. *98*(8), 484–495.

Wong, M., & Evans, D. (2007). Students' conceptual understanding of equivalent fractions. In J. Watson & K. Beswick (Eds.), *Proceedings of the 30th annual conference of the Mathematics Education Research Group of Australasia* (pp. 824–833). MERGA.

CHAPTER 4

MEASUREMENT

Derrick W. Smith and Amanda Lannan

One of the most widely used aspects of mathematics is the concept of measurement. In order to make sense of the world, humans have devised ways to quantify important aspects of our lives. For example, to explain the distance from one object to another, we created the concept of *length*. While discussing a two-dimensional area, we conceived the concept of *area*. To make meaning of the space within three-dimensional spaces, we adapted the concept of area in the form of *volume* and *capacity*. In order to explain how "heavy" something may be, we created standards of *weight*. To measure how much of a substance exists, we created the concept of *mass*. Through our interactions with our environment, we realized we needed a way to discuss *temperature* and thus devised multiple ways to measure it. Due to our need to exchange goods and services, we placed value on a *currency* of trade. And through our need to quantify our own lives, we established a way to measure what many consider our most valuable commodity, *time*. All of these—length, area, volume/capacity, weight/mass, temperature, currency, angle measurement, and time—are the fundamental elements of measurement that we learn and use throughout our lives.

The learning of measurement begins at a very early age as children attempt to make sense of their world in relative terms. The formal learning of measurement typically begins by using nonstandard measurements before moving on to more standardized measurement systems. For example, young children learn the basic concept of relative temperature through the experiences in their environment (e.g., a "hot" stove versus a "cold" ice pop). Throughout their lifetime, they will slowly shift from relative temperatures (warm/hot or cool/cold/freezing) to more exact, standardized measurements (100 °F versus 32

°F versus 0 °F). The role of the educator is to ensure that students acquire a foundational understanding of measurement in such a way that has meaning to their lives and supports their current and future endeavors.

While measurement begins in the early school grades, teachers continually expand upon the concepts as students make progress to more advanced concepts. Measurement, while primarily taught within mathematics, is vital to all academic areas. In science, practitioners use all forms of measurement to quantifiably make comparisons to a standard. Within social studies and English language arts, the concepts of time, currency, and even area are tools that help learners make meaning of the content. Art, music, business education, and even physical education all use most, if not all, of the measurement concepts in some form or another. In truth, measurement is a key tool for one to become truly "numerate." To have the ability to create, analyze, and communicate about data and use that information to make informed decisions, students must have a strong understanding of measurement.

GENERAL CONSIDERATIONS OF TEACHING MEASUREMENT

In thinking about teaching measurement to all students, there are some inherent considerations for all professional educators. First, measurement is typically a focus in the early grades but becomes an ancillary topic in the later grades. In reviewing the *Common Core State Standards in Mathematics* (National Governors Association Center for Best Practices & Council of Chief State School Officers, 2010), grades kindergarten through 5 each have a domain specific to measurement. However, beginning in the sixth grade, we find measurement concepts embedded in other domains such as geometry, statistics, and probability, and in some algebra-related standards. While students must learn a great deal of the foundational concepts of measurement by the sixth grade, they must expand upon and master the concepts throughout the later grades. This unfocused inclusion in the middle and high school curriculum (outside of a standalone geometry course) means that some students may never master the necessary aspects of measurement they need to become truly numerate.

Another consideration in teaching measurement is that the curriculum tends to rely heavily upon the procedures and not on understanding the concepts themselves (Smith, Males, & Gonulates, 2016). As an example, consider how schools often teach the concept of area. After spending adequate time on measuring length using measuring tools, the concept of area shifts from the use of physical tools to numerical computations (formulas). When tasked with determining the area of a polygon, students immediately shift from a dedicated "hands-on" approach for finding length to using a formula to determine the area. While many students may be able to complete the algorithm for determining the area of basic shapes, this does not necessarily mean they have a true understanding of what area really means. For example, they may complete a problem to compute the area of a square as being 16 square inches but never really understand what "square inches" really means. The related challenges may become more apparent when students encounter irregular shapes composed of multiple regular polygons. While the skill of employing formulas to calculate measurements is important, genuine mastery remains incomplete if the student conceives of the formulas as abstractions instead of as models for understanding the world.

Measurement is unique in that it is incredibly useful in a variety of situations. For example, measurement is as important to a house painter's business as it is to an aerospace engineer's work. Both use measurement concepts of length, area, time, and currency (just to name a few) to complete the tasks of their profession. Obviously, they may use different aspects and complexities of measurement for their purposes, but the fact remains that most professions use measurement. Educators thus face the challenge of ensuring that students not only learn the concepts to understand and the procedures needed to compute measurements, but they must see the applications of measurement as well. This means that learning activities must move beyond simple procedural problems and explore the complex uses of measurement across multiple domains.

CONSIDERATIONS WHEN TEACHING MEASUREMENT TO STUDENTS WITH VISUAL IMPAIRMENTS

Due to its inherently visual nature, teaching measurement to students with visual impairments can pose unique and significant challenges. Many fundamental measurement concepts, such as length, area, volume, and capacity, originate from spatial perceptions that heavily rely on sight. From an early age, children learn to distinguish objects by their size, primarily through visual comparisons. While they can still discern larger differences through tactile or visual means, the difficulty escalates as the differences become subtler. The smaller the difference, the more challenging it becomes for a child with visual impairments. For example, a student could compare 1 cm to 6 cm with a simple tactile comparison. However, comparing 1 cm to 1.1 cm would require not only the use of a measurement tool but also the ability to determine the exact differences. While students with visual impairments can still learn these spatial concepts, educators should be ready to adapt the visual elements in a lesson—not only through the use of accessible graphics, but also through accessible measuring tools such as large-print and braille rulers or digital measuring devices that auditorily provide precise measurement information.

Other concepts, such as time and currency, have their own set of challenges related to vision, as the design of widely used systems for understanding and measuring them are not easily accessible by students with visual impairments. For example, students learn to differentiate monetary denominations at an early age, primarily by sight. While coins have many distinct physical characteristics that differentiate them, students with vision tend to focus on the color, size, and imprints on the coins. For students with visual impairments, the size is still important, but they may also use other identifying features, such as size or whether the outer edge of the coin is smooth or ridged. Those clear differentiators do not exist in paper currency, as all bills in the United States are the same size and are only distinguishable by their images and numbers. There are additional potential challenges when it comes to the concept of time for students with visual impairments beyond the visual nature of most time-telling devices. For example, difficulty connecting the time of day with the position of the sun may present difficulties when organizing events within daily routines.

Teaching measurement to students with visual impairments also presents a distinct set of challenges related to the concepts of "scope" or size. In truth, even individuals with vision sometimes struggle with this concept as well. For example, it is one thing to say that a billion is 1,000 million, but understanding the difference between a million and a billion is quite abstract. At a more basic level, scope can be a greater challenge for students with visual impairments, as they do not always have access to large-scale objects for comparison. For example, a class may be working on exploring the length concept of a mile. Therefore, a teacher may take the class outside during recess or p.e. and have them mark a starting point and then walk a mile as a class. Once finished, the teacher can have the students turn around and look for the place where they started. The teacher could then show the students a footlong ruler and a yardstick and explain that a mile is 5,280 feet or 1,760 yards. By looking at the comparison, the students begin to develop the concept of a mile. However, for students with a visual impairment, they would not be able to see the full length of the mile that their peers can easily identify. While they would have walked with them and understood that it is a long way, they would not have the same depth of understanding as do their sighted peers. The same is true when thinking of other large objects such as an office building. Although a teacher can explain to a student that an office building has 15 floors, they likely would not understand the complete scope of what that means other than it is high.

The single most difficult challenge to teaching measurement to students with visual impairments is its heavy reliance upon specialized tools for precise measurements. The first part of this challenge is that measurement uses tools. We have the ruler (in its various forms) for length, area, and volume. We have scales for weight. We have clocks for time and thermometers for temperature. We have protractors to measure angles. While it is possible to adapt most basic measurement tools for accessibility, the challenge of precision appears: Even with an accessible ruler, how challenging is it to measure down to the nearest millimeter? Whether students are using basic tools such as rulers or high-tech devices such as microscopes or cameras, students with visual impairments must develop the skill set needed to utilize all the accessible tools effectively.

These are only a sampling of the many tools that students can use for measurements. As students advance in their studies, particularly in the sciences, they use more advanced instruments to collect measurements. While many of the tools have digital outputs that may be accessible, students would still need specific instruction to ensure that they complete their measurements correctly to acquire the correct outputs.

KEY STRATEGIES FOR TEACHING MEASUREMENT

Measurement is a fundamental numeracy skill that has many uses in life. It is simple to determine its many general uses, but there are many professions, such as engineering, health care, business, trade work, etc., that utilize measurement on a daily basis. While there are unique challenges for teaching measurement concepts to students with visual impairments, there are just as many strategies for overcoming them. Through a combination of specialized, compensatory instruction with accessible tools, students with visual impairments can develop their measurement skills. All students should use the following six "keys" to ensure a strong foundation in measurement.

Key 1: Begin With Nonstandard Units of Measurement

For students to understand the construct of any type of measurement, they should begin by exploring their world using nonstandard measurements. For example, to begin exploring temperature, a teacher may discuss how it is cold outside versus being warm inside. They could discuss how the temperature inside changes due to a heater that warms the air. Another example would be to start developing the concepts of volume and capacity by having students determine how many similar toys it would take to fill up a bucket (thus determining the capacity of the bucket in terms of the toy). As the students explore these physical features, the teacher should have discussions with the students about the need for us to quantify things so all of us can understand the concept. Then, they can explore how nonstandard units are helpful but not exact enough in every situation. For example, one person may say it is cold outside while another person may consider it to be cool. What is the relative difference between "cold" and "cool"? When students understand that the difference between the two is a matter of perspective, they will recognize the necessity of strategies that rely on specific measurements.

Key 2: Shift to Standardized Units

At some point, the students must come to the realization that measuring things using nonstandard measures is inconsistent. The structure of the lessons should be for students to construct a need for standardized measures. Once the students have established the need, the lessons should then center on using standardized units of measurement to quantify real-world objects. In other words, have the students measure things!

Key 3: Focus on Situation-Appropriate Precision

Having the students use measurement tools will encourage them to measure with precision. The educator should require students to ensure that their measures are precise when it is appropriate. For example, if a person is attempting to determine the exact dimensions of a space to ensure that an object will fit, then precision is important. There are many situations in the real world where precision is absolute.

Key 4: Connect Precision to Approximation

However, there are many times when precision is not necessary, and approximation is sufficient. Teachers in the United States tend to teach students to focus on precision more than anything, when in reality approximation is an incredibly useful skill. The teacher should focus on helping the students to develop a "sense" of the units of measurement (such as generally how long an inch is or how much 5 pounds weighs) and then apply that to various situations. Generally speaking, there are many situations when approximations are appropriate for solving problems that involve measurement. For example, if I were making pasta, how much water would I need to boil a pound of dry pasta? The instructions on the packaging may tell me to use exactly 4 quarts. Therefore, I can measure out exactly 4 quarts or approximate 4 quarts by adding or pouring out some water I already have in the pot. Either way, I will need to pay attention to the water as I cook the pasta. A few ounces of water will probably not make a major difference in making the pasta.

Key 5: Use Measurement in Authentic Assignments

In other words, have the students use the measurement concepts they learn in real-world situations in their own lives. While there is a time and place for learning how to measure and then use the measurements to compute specific measurement quantities (typically by using formulas), it is also critically important for students to connect that knowledge and those skills to situations that have meaning in their lives. Contrived word problems from the textbook are a good way to practice concepts, but working with the students to develop problems (and solutions) in their own lives is always a better way to learn.

Key 6: Teach Both Standard and Metric Systems

While there is some resistance to teaching both systems of measurement in U.S. classrooms, the reality is that students in the United States must be adept at using both U.S. Standard System of Measures ("standard") and the

Metric System of Measures ("metric"). Within the United States, the most common measures used in everyday life are the standard measures (inch, foot, yard, pound, ounce, etc.) and so students must be able to use these measures effectively. At the same time, the metric system is the predominate standard of measurement used around the world and is the only standard of measures used within the scientific community. Thus, to provide our students with the most career choice opportunities, they need to be able to use both the standard and metric systems. Therefore, we must teach both, often at the same time. However, simply teaching our students how to measure using both systems is not enough; they must also learn how to complete conversions within and across each system as well.

THE PILLARS OF MEASUREMENT

Length

Simply put, *length* is a measure of distance. It is the measurement or extent of something from end to end. Within geometry, the definition of length is the straight-line distance between two points along an object (Weisstein, 1996). Various terms connected to the concept of length include *height*, *width*, *breadth*, or *depth*. These various terms, while typically used in describing lengths in different directions within a three-dimensional space, are all "lengths."

Within the standard system, inches are the foundational unit of measures and used to measure length. For longer lengths, combinations of inches make up the foot, yard, and mile. Fractional parts of the inch measure increments typically down to the 32nd part of an inch (meaning $\frac{1}{32}$ of an inch). For the metric system, the base unit is the meter, and its different lengths are all powers of 10. Whether greater or smaller than the meter, powers of 10 serve as the base for different measurements and different prefixes assign meaning (a centimeter is 10^{-2} or 0.01 of a meter, while a kilometer is 10^3 or 1,000 meters).

Teaching Length to Students With Visual Impairments

As noted within the keys to teaching measurement, the teacher will want to begin teaching length using nonstandard measures. Most of these activities are accessible, as students can use their hands or fingers or objects (such as paper clips or straws) to "measure" an object (textbook, desktop, etc.) using the item. Activities that students can complete at the desk are typically accessible with minimal modifications. For example, if the activity calls for paper clips, students may need to interlock the clips so they are easier to lay out together. At the same time, the teacher may want to place a nonslip surface mat on the object to help keep the paper clips from sliding off while the students are measuring. For activities that are larger in scale, the teachers may need to be more thoughtful about how to include students with visual impairments in the exercise. For instance, what if a teacher decided to ask the students to measure the length of the classroom using a student's body length? In this situation, it may be best for a student with visual impairments to serve as the unit of measure. Have them lie on the floor, put down a marker, and then have them shift to the next section and so on. The other students could make the marks, count the marks, and help add up the count while the student with visual impairments has a tactual experience that strengthens their understanding.

For lengths beyond a yard or meter, a teacher may have the students go outside and walk a specific distance so they can "see" a specific distance. While such an activity includes some challenges for students with visual impairments, they could participate in the walk to get a general idea of the length. Maybe they could count their steps as they walk to help measure the length. Alternatively, if the class has access to a football field or another open area, students could stand at various distances away from the starting location, then call out their distance or clap to provide an audio reference. For developing the concept of distance as it relates to time and speed, working with an orientation and mobility (O&M) specialist would be appropriate for students, as developing a sense of distance and the time it takes to cover a distance (speed) is part of their learning objectives. Learning to standardize the general distances between doorways in a building or the distance of mailboxes from one another in a neighborhood are all applicable ways to develop a sense of length covered in an O&M lesson.

For very small lengths, the greatest challenge for students with visual impairments will be the limitations of precision at very small measures. When using accessible tools such as the Braille Ruler With Caliper from the American Printing House for the Blind (APH; see Figure 4.1) or the Tactile Caliper sold by National Braille Press (see Figure 4.2), students with visual impairments must learn to use a caliper to hold the ruler in place for precise measurements. The APH ruler (available at aph.org/product/ruler-1-foot-braille-metric-english-measurement) requires students to hold the caliper and read either the large print or braille on the ruler. The Tactile Caliper (https://shop.nbp.org/products/tactile-caliper) also requires the user to read braille, but it provides the reader with the precise measurement increment as a fraction down to the nearest $\frac{1}{16}$ of an inch (standard) or 1 mm (metric). However, students may still require support from others for any measurement that is more precise than these, or the assignment may permit them to measure to within plus or minus a predetermined error of measure.

Figure 4.1 1-Foot Braille Metric/English Ruler With Caliper

When measuring with a braille ruler, students will need to use the caliper to ensure a precise reading.

Figure 4.2 Tactile Caliper

A tactile caliper like the one from National Braille Press pictured here works well for measuring discrete objects.

As noted, it is possible to describe length in other ways in relation to directionality in terms of width, breadth, height, and depth. In working with three-dimensional shapes, the students should explore the different lengths of the sides in correlation to how they can commonly use these terms. For example, width is typically the measure of a horizontal line moving side to side, and breadth is the same but for especially large spans of distance. Depth typically refers to the distance from the line used for width and another line parallel to it on the same horizontal plane. Height is typically the vertical length from one end of the width line to another parallel line above or below on the vertical plane. While this is all quite technical in nature, teachers can impart these concepts easily, and students can understand them with a box and by spending some time exploring the line segments that form the edges of the box.

Perimeter

Length relates to the general concept of *perimeter*. The definition of perimeter is the distance of a closed pathway that outlines a two-dimensional shape. To compute perimeter, add up all the lengths of the line segments that outline the shape. (Note that the perimeter of a circle is the *circumference*.) As part of the learning process, students should have opportunities to not just memorize the basic formulas for general shapes but also derive those formulas through exploration. For example, if given a triangle of any type, students can determine that they can calculate the perimeter by merely adding the three sides. However, when given a rectangle, the students should notice that the figure has two sets of sides with the same scaffolded lengths to construct the formula $P = 2l + 2w$ or $2(l + w)$. It is essential that the students understand the reasoning behind these formulas so that they truly understand the concept of perimeter and not just the formulas themselves. Often, many students struggle to make meaning of the concept of perimeter and only think of it as a formula.

Area

Directly connected to length is the concept of *area*. The definition of area is the measurement of the size of a two-dimensional surface. In its most basic conception, area is the number of "squares" with a specific dimension (such as 1 inch by 1 inch, equivalent to 2.54 cm by 2.54 cm) needed to cover the surface of an area. As an example, one can compute the area of a rectangular room's floor by multiplying the two lengths of the two sides (length × width) in a unit of measure. If the answer is 20 feet squared, then 20 1 × 1-foot squares would cover the entire floor. *Surface area* is a connected concept in that it is the total area of a three-dimensional object's surface. It is simple to compute this by decomposing the three-dimensional shape into multiple two-dimensional shapes, determining the area of each individual shape, and then adding all the areas together.

Within the standard system, people most often measure area in *squared* units (such as squared feet or ft^2). The only other commonly used measure of area that involves area is an *acre*, which is 43,560 square feet (equivalent to 4,046.9 square meters). However, within the metric system, one can use any unit squared, most commonly the measure of square hectometers/hectares (10,000 square meters, equivalent to 107,639.1 square feet) or square kilometers in reference to large areas.

Teaching Area to Students With Visual Impairments

In order to master the concept of area, students must first understand what area actually means. A common misstep in the early grades is when the instructional focus seemingly "jumps" directly from measuring lengths to using formulas to compute areas and more advanced measures. High-quality instruction about area begins by exploring the meaning of the measure through hands-on experiences. For example, a teacher may ask students to determine how many Post-It Notes they will need to cover the entire space of their desk or textbook, or the teacher may have the class work collectively to determine how many tiles it would take to cover the entire classroom. These types of hands-on experiences are easily and simply adaptable for use by students with visual impairments.

Another common activity is to have students use grid paper to construct polygons and then count the squares that make up the area of the polygon. This may include having the students create the boundaries of the polygon and then even color the squares as they count them. Students with low vision could use dark-lined graph paper to complete these tasks. For students who read braille, using APH's Embossed Graph Sheets (Figure 4.3; aph.org/search-results/?fwp_search_term=embossed+braille+sheets) and something to mark the boundaries would be all they need. For example, a teacher could provide the students with Wikki Stix (wikkistix.com) and have them create the boundaries of the polygon in the same way as their classmates. Then, the students would need to systematically count (such as starting in the top left-hand corner and counting from left to right) the number of squares within the boundaries. Another option would be using APH's Graphic Aid for Mathematics (Figure 4.4; aph.org/product/graphic-aid-for-mathematics-2) to complete the same task.

Figure 4.3 Embossed Graph Sheets (Detail)

Embossed graph sheets allow braille readers to access activities in many areas of mathematics.

Figure 4.4 Graphic Aid for Mathematics

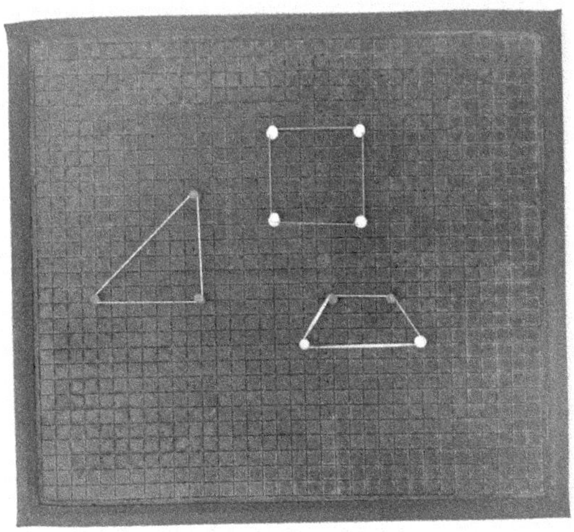

The Graphic Aid for Mathematics is a flexible tool for math instruction.

The next step in the learning process would be to shift from manual counting using manipulatives to deriving the formulas starting with the square and rectangle. If the class uses graphics as they derive the formulas, introducing accessible manipulatives and/or high-quality tactile graphics will support the instruction. The area of triangles typically begins with cutting a rectangle in half., (Using shapes made of cardstock paper or foam paper may be best for manipulation.)

To explore surface area, APH's Geometro shapes (aph.org/search-results/?fwp_search_term=geometro) provide an excellent tool for teaching three-dimensional shapes (i.e., polyhedrons). Before teaching the formulas, students need to develop a basic understanding that they can determine the surface area of a three-dimensional shape by adding together the areas of the polygonal sides of the polyhedron. By using the Geometro (Figure 4.5) or other commercially available polyhedral "nets" such as Omnifix Cubes (didax.com/omnifix-cubes-set-of-100.html), students can use their understanding of area to derive the surface area formulas by combining the area formulas algebraically.

Figure 4.5 Geometro Shapes

Geometro kits are ideal for teaching the relationship between 2-dimensional shapes and 3-dimensional polyhedrons.

Volume and Capacity

Volume is the total space occupied by a three-dimensional object, whether solid or hollow. Most people typically measure volume numerically in cubic meters and its derivatives (such as cubic centimeters or cubic millimeters). The primary concept of volume is that it tells us how many "cubes" of any single dimension would take up the same space as the object. For example, a cardboard box (cube shaped) may measure 1,000 cubic centimeters. Thus, we would be able to construct another cube of the exact same size out of 1,000 cubes with dimensions of 1 cm by 1 cm by 1 cm. Within the standard system, discussion of volume in general uses measures of "cubic" lengths (such as cubic feet, ft^3). Similarly, the metric system measures volume using cubic lengths, such as cubic meters (m^3), as its units of measurement.

Capacity, on the other hand, is the maximum amount of matter an object can hold and is primarily related to hollow objects. Various situations, from chemistry to cooking, use this concept; for example, a person (such as a baker) may fill up a three-dimensional object to a specific capacity (such as a cup filled with flour). An intimate connection exists between volume and capacity, as the capacity of a container generally defines the volume of an object. Solid objects have a volume, but they do not have a capacity. The standard system measures capacity using measures for fluid volume, such as fluid ounces, cups, pints, quarts, and gallons. On the other hand, the metric system uses liters as its base unit for capacity.

Teaching Volume and Capacity to Students With Visual Impairments

Too often, the study of volume is relegated to merely learning the formulas. However, high-quality instruction will begin by having students explore how to compute volume using tangible manipulatives. For example, using stackable cubes such as APH's StackUps (Figure 4.6; aph.org/product/stackups-kit-spatial-reasoning-using-cubes-and-isometric-drawings) or Didax's Omnifix Cubes, a teacher could have students create a cube or a rectangular prism. By creating these three-dimensional shapes using small "unit" cubes, the students can explore the length, width, and height of the figure and count the volume per unit cube. From this, the teacher can then lead students through the process of deriving the formula for the volume of a rectangular prism ($V = lwh$).

Other three-dimensional shapes are not as simple to create with smaller cubes compared to a cube or a rectangular prism. However, from the rectangular prism example, the teacher can explain that the formula for the volume of three-dimensional shapes is the area of the base (length × width) multiplied by the third dimension (typically height). With more advanced shapes that incorporate curvatures (such as a cone or sphere), the explanation requires an understanding of calculus that teachers often do not explain in depth in high school. For a student with visual impairments, the explanation of the calculus problem will need to follow the one all the students receive, all while incorporating three-dimensional manipulatives and tactile graphics.

Figure 4.6 StackUps Kit

The Stackups Kit and similar manipulatives have small "unit" cubes that can be used to teach volume.

Capacity is a related concept that students may or may not learn along with volume. In the early grades, instruction will focus on teaching students how to measure fluid and some solids (such as sand) using capacity measures (ounces, cups, pints, quarts, and gallons). These lessons can stand alone or connect to the expanded core curriculum (ECC) concepts such as cooking. There are many commercially available measuring tools (including items such as teaspoons and tablespoons) that have large print and different color schemes such as yellow print on black backgrounds for students with low vision. Braille measuring tools are also available from many resources that specialize in adapted independent living aids. Students with significant vision loss can also use standard kitchen equipment such as measuring spoons and cups (which they can fill and then tactually remove the excess from), but some tools such as measuring glasses are challenging to make accessible. A strategy would be to use accessible measuring tools to fill a larger measuring cup or glass. The instructor could line the inside of measuring cups with tactile tape (such as striping tape used for drafting) so that the student can feel when the material reaches the measurement mark. Of course, the roll of tape would need to be kept clean if students are using the measuring tool for cooking, and the material used to fill the cups would need to be something that they can touch.

Weight and Mass

In simple scientific terms, *weight* is the amount of gravitational force or acceleration exerted on an object, while the related but distinct *mass* is the amount of matter within an object. Mass is always constant, but weight is not. For example, the lunar rover would have the same mass on Earth and on the moon, yet it would weigh less on the moon due to the reduced gravitational force exerted upon it. Connected to the concept of mass is the idea of *density*, which is the measure of a substance's mass per unit of volume. Thus, the higher the mass-to-volume ratio, the greater the density, and vice versa. As an example, one could take two cars: one real and one made of foam. Due to the differences in their density, their total mass and total weight would be different. The standard system uses ounces, pounds, and tons as their primary units for weight and mass, while the metric system uses grams as its base unit.

Teaching Weight and Mass to Students With Visual Impairments

While weight and mass are distinct measures, schools often teach students about them together, primarily in science classes. However, teachers can convey the essential elements of both measures mathematically as well, particularly as students learn to compute different measures (such as 16 ounces weighing 1 pound). Regardless of the type of classroom setting, students should learn about weight initially using relative comparisons. Instruction typically begins with exploring comparisons of weights and denoting them as "heavy" or "light." As students' understanding of weight progresses, they can connect weight to standardized measures. There is a variety of scales that students may use to measure weight, including digital scales and spring balance scales (such as what may be seen in a produce market). There are many accessible, talking digital scales on the market that teachers can use to teach students weight concepts. The APH MathBuilders: Measurement kit (aph.org/product/mathbuilders-measurement) includes a balance scale that students can fill to determine when the scale is even. Students can also use accessible weights with large print and braille to measure the weight of objects.

To develop a clear understanding of mass, students should have opportunities to investigate using "balance scales." There is a variety of balancing scales they can use, from very basic types where the sides must be equal to more advanced scales that provide digital output. Some balance scales are more accessible than others for students with visual impairments, so teachers will need to select one that can provide tactile feedback at the fulcrum without impacting the balance.

Using scales of any type to compare objects' weight and mass is the best way to explore the concept. For students with visual impairments, instruction that focuses on developing a general understanding of weight and mass of common objects can help them in many areas of their life. Understanding the differences between 1 pound versus 5 pounds can help in areas such as cooking, shopping, and health. Manufacturers and pharmacists, among other professionals, measure many medications in units of mass such as milligrams (mg). Scientific fields from medicine (healthcare) to engineering to pure science (biology, chemistry, physics, etc.) use weight and mass for their calculations. Beyond their uses in professions, many aspects of everyday life employ weight and mass measurements.

Temperature

Temperature is a physical quantity to measure the perceptions of heat and cold. There are three major common scales used to measure temperature: *Celsius* (°C), *Fahrenheit* (°F), and *Kelvin* (K). People in the United States primarily use Fahrenheit while most other countries use Celsius. In scientific purposes, Kelvin is the primary scale. Celsius uses 0 °C as the freezing point of water and 100 °C as the boiling point of water. Fahrenheit uses 32 °F as the freezing point of water and 212 °F as the boiling point of water. Kelvin is a much more advanced scale used for extreme temperatures primarily found in scientific endeavors; it is beyond the scope of this text.

Teaching Temperature to Students With Visual Impairments

The scope and sequence for teaching temperature to all students begins by helping them understand relative temperature. In the early grades, discussions center around concepts such as "hot" and "cold" and relate those air temperature and other common examples such as "ice is cold" and "fire is hot." The instruction of relative temperature continues naturally with the use of different levels of temperature such as "cool versus cold versus freezing" or "warm versus hot versus scalding." Students with visual impairments have a similar sense of air temperature to their peers' and are able to engage in these early learning activities with limited accommodations. Teachers can make many of the learning activities related to relative temperature both accessible and safe for all students with minimal accommodations. However, take caution and remember that relative temperature does have extremes wherein it becomes unsafe to test by touch.

Figure 4.7 Tactile Demonstration Thermometer

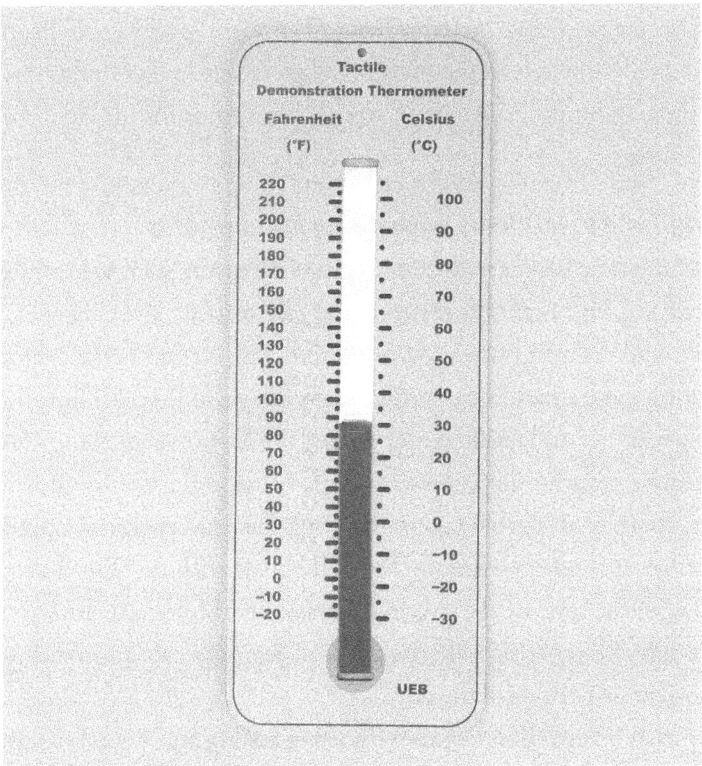

The Tactile Demonstration Thermometer shows the relationship between Fahrenheit and Celsius.

As the sequence of learning shifts to standardized temperatures, students typically begin to relate relative temperatures (freezing) to standard temperatures (32 °F or 0 °C). To make the learning activities accessible to students with visual impairments, use of a thermometer that has audio output (i.e., "talking thermometer') in the classroom is indispensable. For example, one common class activity is to have the students read a thermometer each day to tell the class what the air temperature is outside before going to recess. The teacher could use the APH Tactile Demonstration Thermometer (Figure 4.7; aph.org/product/tactile-demonstration-thermometer-ueb-nemeth) to show the temperature in Celsius and Fahrenheit to the entire class in both large print and braille.

Time

Time is the measure of three things: the concepts of duration or elapsed time, sequences of events, and the beginning and ending of events. Time is a continuous measure that includes the past, present, and future. People measure time predominately in the form of progressively longer durations, including seconds, minutes, hours, days, weeks, years, months, decades, centuries, and millennia.

Teaching Time to Students With Visual Impairments

Teaching time begins with a focus on relative time such as morning, afternoon, evening, and night. Students begin to construct the concepts through repetition and discussions related to the sun and events that commonly take place during those times. For example, many early childhood classrooms have a "morning routine," and teachers may discuss the "at-home morning," "evening," and "bedtime" routines. For students with visual impairments, most of these activities are fully accessible except for using the sun's position to understand relative time. It is not uncommon for individuals with no light perception to struggle with the time of the day and setting their biological clock to the day. But through the repetition of routines, all learners can construct meaning from the general periods of the day.

Very early, instruction begins with using analog and digital clocks. In the early grades, teachers will begin to connect a specific time with an activity (such as, "It's 11:30 a.m., so it is time for lunch!"). While digital clocks are easier to read for everyone, teachers and parents still teach about the analog clock, as it has connections to geometry and algebra as well as to orientation and mobility. For example, people may present orientation information using the "face" of a clock, such as "The chair is at 12 o'clock, and the board is at 4 o'clock."

An array of analog clock tools and manipulatives are available to aid in teaching this concept. For example, it is not uncommon for teachers to have a class set of "play clocks" to support the explanation of hours and minutes. For students with low vision, some of these clocks may have large enough print on them for them to be useful. However, APH also offers an accessible Analog Clock Model (Figure 4.8; aph.org/product/analog-clock-model) that includes both large-print and braille numerals.

Figure 4.8 Analog Clock Model

A model of a clock can help students understand how people measure time.

Teachers often use graphics to teach and assess students' ability to read analog and digital clocks. As with any graphic, the use of enlarged graphics for students with low vision and tactile graphics for students with limited functional vision is essential. APH has Clock Face Sheets in Braille (Figure 4.9; aph.org/product/clock-face-sheets-in-braille) that are useful for direct instruction or assessment. Using brass fasteners, Wikki Stix, or string can help support instruction of the analog clock. Instruction typically begins with the hour hand and then shifts to the minute and second hands. There are numerous accessible analog watches and clocks on the market that students can use. However, students must also understand how to effectively use a digital clock or watch. There are many accessible digital clocks available through a variety of technologies (standalone clocks, smartphones, braille devices, etc.). APH's Flip-Over Concept Books (aph.org/search-results/?fwp_search_term=-flip-over+books) provide an excellent tool to explore both analog and digital clocks and connect the two concepts.

Figure 4.9 Clock Face Sheets in Braille

Braille clock faces help with assessment of concept mastery.

Of course, time is more than just reading a clock—it also involves an understanding of greater lengths of time, such as the day, week, month, year, etc. In the earliest of grades, students will explore using calendars to develop their understanding of the names of days, how those days construct weeks, and then how those weeks form months. "Calendar work" is a critical step in the understanding of time and has many applications in education (such as in studying history) and our lives. All students should have access to calendars in a medium they can read. For students with low vision, a variety of software tools can help with the creation of high-contrast calendars in the appropriate color scheme and font type and size. For students who read braille, APH's Classroom Calendar (aph.org/search-results/?fwp_search_term=classroom%20calendar) and the Individual Calendar Kit (aph.org/product/individual-calendar-kit) are excellent tools. The Classroom Calendar is completely accessible for all students, including the entire class and not only one student (or more) with visual impairments.

For students with multiple disabilities and visual impairments, instruction on time may take the form of creating a daily routine or a task analysis using divided boxes. There are many available "divided boxes" on the market, they can use to create these routines. Connect the routines to time as appropriate for the student. The items within the routine boxes should be appropriate for the ability of the students to connect an item to a task or to a time of the day.

Currency

Currency, or *money*, is any item used as a medium of exchange that a society generally accepts as payment for goods and services. Currency is the standardization of money in any form and typically exists in the form of banknotes (paper money) and coins. Of all the types of measurement discussed, currency is the only type that is typically unique to a specific country or region. While there is some larger scale standardization such as the euro (used in most European countries), many countries have their own established forms of currency.

Teaching Currency to Students With Visual Impairments

From an early age, students often have an understanding that the purpose of currency, or money, is to purchase items or services. They often learn this by watching adults exchange money for things. For students with visual impairments, there may be a lack of incidental learning in which they may not fully understand what it means to "buy" something. Therefore, it is important that teachers, including O&M specialists, explain, model, and allow students with visual impairments to use currency to understand "how it works." In our modern era with the use of debit and credit cards, many students may not have a clear understanding of how purchasing occurs, so it is incumbent on teachers to help students better understand how monetary systems work.

While some students may understand that they may use money to purchase items or services, they often do not have a concept of the value of money. For instance, a young child may be more than willing to take two dimes for one quarter because they think that "2 is larger than 1." Developing "money sense" is a primary focus not only in early elementary grades but throughout formal education.

Students begin to develop money sense through learning the value of coins and comparing them to the base unit of money, the penny. While teachers often use "fake money" to teach these concepts, for students with visual impairments, it is critically important to use real coins so that the instruction can focus on the distinct features of each coin. While students with vision primarily focus on the size and appearance of the coins, students with visual impairments must receive instruction on differentiating them by size, weight, and texture on the edges.

Paper currency is much more of a challenge for students who use braille, yet with the correct tools even discriminating currency is possible. Individuals who are blind or visually impaired can get a free currency reader called iBill from the U.S. Bureau of Engraving and Printing (bep.gov/services/currency-accessibility/us-currency-reader-program). With this device, the user inserts a U.S. bill into the reader and presses the button, and the device will state the value of the note (see Figure 4.10). There is a variety of apps for smartphones that allow a user to take a picture of a paper bill and will tell the individual the denomination of the bills. Professionals who work with students with visual impairments should help their students explore using these technologies as part of the curriculum. Students should also learn how to organize paper currency in their wallets. One strategy is to keep all $1 bills flat, $5 bills folded in half vertically, $10 bills folded in half horizontally, and place anything larger in separate compartments in the wallet.

Figure 4.10 iBill Currency Reader

The iBill Currency Reader is available for free from the U.S. Bureau of Engraving and Printing.

As part of the curriculum related to currency, students should engage in other monetary-focused activities including using cards to pay for purchases, writing checks, using mobile payment apps, budgeting, financial planning, taxes, etc. While the general curriculum may cover most of these skills, the task of ensuring that students learn these independent living skills may fall upon the teacher of students with visual impairments (TSVI) as part of the ECC with support from the O&M specialist. There are multiple check-writing guides for students with limited vision to use to help write out checks. (An example is at aph.org/product/keitzer-check-writing-guide-for-veterans.) Learning how to sign their names with and without a signature guide is also a skill that should be part of the currency curriculum. As with all these advanced topics related to finance, the goal would be for the students to have access to the materials in such a way that they can learn the concepts.

Angle Measurement

Angle measurement is a type of measurement used to measure the distance between the two rays that form an angle. Either degrees or radians are the units used to measure the space between the two rays, called the vertex. With a circle divided into 360 equal degrees, the measurement of angles typically falls between 0 and 360 degrees, depending on the distance from the initial ray and the terminating ray. In general, the degree (denoted with the ° symbol) is the primary unit of angle measurement used in geometry, trigonometry, algebra, and calculus. However, to allow for more precision in advanced math, one can divide the degree into 60 equal parts called minutes, denoted with a single apostrophe. One can further divide each minute into 60 equal parts called seconds, denoted by a quotation mark. Therefore, a written example of 5 degrees, 25 minutes, 45 seconds would be 5° 25' 45".

A second common measurement for angles, used primarily in algebra and calculus, is the radian. While the degree focuses on measuring the vertex of an angle by dividing a circle into 360°, the radian utilizes the mathematical relationship between the circumference of a circle and the radius ($C = 2\pi r$) and divides the circle by 2π. There is a direct correlation between degrees and radians, as 1 radian = $360°/2\pi$ or $180°/\pi$.

Teaching Angle Measurement to Students With Visual Impairments

Students typically receive an introduction to angle measurement in around the fourth grade in relation to geometry. As noted in Chapter 5, "Geometry," all students should receive instruction on how to use a protractor to measure angles. Students with low vision should use angles with dark lines coupled with a dark-print or large-print protractor; for students who use braille, teachers should provide the angles as tactile graphics and also supply a braille protractor. For any student with an iOS device, they may use the free APH Draw2Measure Protractor app (aph.org/product/draw2measure-protractor-app-for-ios-devices-free-app) that utilizes the built-in gyroscope sensor within the device to measure an angle. The Draw2Measure Protractor app provides measurements in both exact degrees and radians.

Students typically receive an introduction to radians in advanced secondary math courses including trigonometry, algebra, and calculus. Radians are typically introduced with the use of the "unit circle," a circle with a radius equal to 1 and the vertex that originates from the point (0, 0) on a coordinate plane. Using the unit circle allows trigonometry functions on the same scale and promotes consistent algebraic computation. For students with visual impairments, it is important that students receive the unit circle (with its labels) in all accessible formats (large print and tactile). APH's Clock Face Sheets are useful for introductory instruction of degrees and radians, but TSVIs may need to create specifically accessible graphics for instruction. Beyond understanding the relationship between degrees and radians, there are no unique accommodations regarding their uses other than learning the mathematical syntax.

CONCLUSION

Measurement is one of the most tangible mathematical skills that students will learn. Individuals need fundamental skills for measuring everything—length, time, temperature, weight/mass, currency—to be able to truly "numerate." While many professions (such as carpentry) use measurement skills extensively, most people use some form of measurement daily, particularly those connected to time, temperature, and currency. It is for this reason that all students should receive access to instruction related to measurement. For students with visual impairments, this instruction may require the use of specialized measuring tools and expansion of instruction, as noted in this chapter. The TSVI and the general education mathematics teacher must collaborate to ensure that the students not only have access to the curriculum but also develop the necessary understanding of the concepts that they will use throughout their lifetime.

REFERENCES

Common Core State Standards Initiative (2010). *Common core state standards for mathematics*. https://corestandards.org/wp-content/uploads/2023/09/Math_Standards1.pdf

Smith, J. P., Males, L., & Gonulates, F. (2016). Conceptual limitations in curricular presentations of area measurement: One nation's challenges. *Mathematical Thinking and Learning, 18*(4), 239–270. https://doi.org/10.1080/10986065.2016.1219930

United Nations. (n.d.). *UN operational rates of exchange*. Retrieved September 2, 2024, from https://treasury.un.org/operationalrates/OperationalRates.php

Weisstein, E. W. (1996). *Length*. Wolfram Research. https://scienceworld.wolfram.com/physics/Length.html

CHAPTER 5

GEOMETRY

Sean Tikkun

In the field of mathematics, there are a number of domains. Just like science has biology, chemistry, and physics, mathematics has algebra, trigonometry, calculus, and many others. Geometry is one of the few areas in mathematics that students learn and identify by name. The word *geometry*, meaning "measurement of the earth," describes the area of study's original goals; in the 20th century, it was synonymous with logic and proofs, but today's students use it principally for applied algebra.

Schools teach geometry to ensure our students can measure, compare, evaluate, and prove information about two-dimensional (2-D) and three-dimensional (3-D) shapes, lines, and angles, but it has a direct connection to how students with visual impairments learn to navigate the world. Geometry provides language and concepts for children as they learn orientation and mobility (O&M) as well. "Parallel traffic," "perpendicular streets," and "adjacent sidewalks" are all terms a student might hear in an O&M lesson that have direct connections to the field of geometry. The concepts in geometry also provide those children who are blind with concepts that assist in understanding the described world. Because of this, it is vital that teachers provide students with visual impairments a solid foundation in geometry that will prepare them for life inside and outside of the classroom.

This chapter proceeds through geometry with increasing complexity and follows the traditional order of Euclidean geometry. Teachers can read through the chapter to gain an overarching primer on the concepts and challenges in geometry, but the sections also serve as a "just in time" review for a teacher of students with visual impairments (TSVI) as they help their students enter a new section or topic. Most concepts in geometry build on foundations that students have had in earlier grades; the sections on regular polyhedra and transformations at the end of this chapter, for example, offer activities that teachers can introduce to preschool students. The chapter provides accessible tools and approaches throughout, but the chapter is not exhaustive in this respect, and teachers should not view the methods shared here as preferential. Students with visual impairments are as unique as their peers and may require varied approaches. (A wealth of additional tools and guidance are available through the Project Inspire course on geometry, available at pathstoliteracy.org/project-inspire-geometry-and-tactile-graphics-students-grades-3-8-course-4.)

PRINCIPLES OF ADAPTING GEOMETRY FOR STUDENTS WHO ARE BLIND OR VISUALLY IMPAIRED

Many of the methods for adapting geometry for students with visual impairments will be familiar to students and TSVIs by the time the students enter a dedicated geometry class in high school. However, as geometry is typically taught with a great deal of visual material, it is important that the mathematics teacher and the TSVI collaborate early and often to ensure students have full access to the materials. The first critical step for collaboration is planning ahead and understanding the accommodative tools available. For example, a general education teacher accustomed to drawing shapes, points, and lines on a board in "real time" as they explain topics may need to rethink their strategy and shift to providing prepared examples alongside manipulatives and tactile graphics. This may also mean that they may need to rely more on the graphics found in the textbooks, which should be already available in accessible versions. The TSVI will need to be more purposeful in introducing tools that are available to make the curriculum accessible. For example, in adapting the materials, TSVIs can use tools such as the Graphic Aid for Mathematics (aph.org/product/graphic-aid-for-mathematics-2) or a Draftsman Tactile Drawing Board (aph.org/product/draftsman-tactile-drawing-board), both available from the American Printing House for the Blind (APH), to quickly create examples or demonstrate concept mastery.

While most geometry instruction relies heavily upon printed graphics, the use of manipulatives can often help with developing many geometry concepts. In truth, the use of manipulatives can be a benefit to all students, not just students with visual impairments. The math teacher and TSVI must collaborate on using appropriate manipulatives that are not only accessible to students with visual impairments but may support instruction for all students. For example, in demonstrating congruence, asking students to demonstrate the principle of reflection by having them flip a physical triangle over a length of string can help them understand that reflected shapes still have identical characteristics. While making this activity accessible for students with visual impairments, it also is applicable to all students.

Contemporary school-age geometry requires a precision and exactness that students with visual impairments may not be used to. Traditionally, this precision was very careful to identify what can be assumed or extrapolated from a graph, drawing from the two foundational texts on the subject (Heiberg, 1908/1956; Hilbert, 1902/1950). It is important to ensure that any adaptations do not create misapprehension in the students or provide information that subverts the purpose of the lesson. For example, there are aspects of geometry that require exact measurement and so may challenge students with visual impairments. However, the TSVI can provide tools that provide as much accessibility as feasible with an understanding that precision may not be possible. The same is true with graphics and labeled measurements. When creating tactile graphics, the labels may need to be put in a different place than they are in print. This must be done with great consideration to not overcomplicate the graphic and possibly impact understanding.

Geometry class is also typically the first time students will be required to use algebra as an application within a different, yet connected, content domain. This applied algebra can be a challenge for any student, not just students with visual impairments. As with all mathematics, the concepts of geometry are additive—they build on one another—so failure to master one aspect of geometry can compound difficulties in the future. In truth, the concepts they learn in geometry are the building blocks for Algebra 2, Statistics, and Calculus. TSVIs should talk to the mathematics teacher to ensure that their adaptations accurately express the purpose of the lesson, and mathematics teachers should be ready to provide student-focused lessons that ensure mastery. These focused efforts are best conducted with the content teacher guiding the need and the TSVI developing the access. For that to happen, a clear, complete understanding of the basic conceptual knowledge required in geometry is key.

BASIC TERMS AND CONCEPTS

Geometry uses language that is commonly known and understood but has strict meaning in mathematics, so it important for the students, math teacher, and TSVI to work from a common, precise language, particularly for students who rely more on instructor descriptions to understand a concept. Teachers should introduce, build up, and use this language consistently during formal geometry instruction to create a lexicon that students will understand in the specific context of math.

Points

The language of geometry starts with the *point*, commonly drawn by making a dot. The dot will be a familiar concept for those students who need tactile materials, thanks to braille. The point has no length, width, or size—it is the fundamental building block for geometry. Points always feature a single capital letter for identification, and most geometric shapes and objects then receive their "names" from the points that define them. For students with visual impairments, it is important that they are provided accessible materials that represent the point, both labeled and unlabeled. For tactile readers, they must become accustomed to locating the point on a page and then searching for the label when provided. Points can be part of other geometry objects and should be distinct enough to be easily found in the graphics.

Lines

A *line* forms when two points connect. Lines come in three varieties: A *line segment* has two clearly defined endpoints; a *ray* has one endpoint and continues indefinitely in a direction; and a *line* continues indefinitely in both directions. The difference among the three is often confusing for students, especially when teachers are imprecise in their terminology. Graphical representations of each of these types of lines commonly involve the line itself, and sometimes points or arrows or both. For students with low vision, the lines should all be in bold, with any points or labels in their appropriate places. For tactile readers, the tactile graphics should include all parts of the line, including the points and arrows. Any labeling should be in braille.

Most of the time when doing geometry-related activities, students work with line segments, which they can identify by the two letters defining the points with a line over them (e.g., $\overline{AB}$). Rays use the same naming convention, albeit with an arrow pointing right instead of a line (e.g., $\overrightarrow{AB}$). This tells the reader that the ray starts at one point and extends through the second point into infinity. (In physics and engineering, vectors also employ a similar naming and appearance practice, but this is based in the mathematical field of vector analysis and not geometry.) The line is infinite in both directions and travels into an imaginary endlessness, which is why the notation includes a double-sided arrow above the letters: $\overleftrightarrow{AB}$.

Figure 5.1 Line Segment, Ray, and Line

It is important for students to maintain the distinction between a line segment, a ray, and a line.

Sidebar 5.1 Basic Symbols for Lines

Line Segment

Symbol in Print: $\overline{AB}$

Nemeth Code: ⠁⠠⠁⠠⠃⠣⠱⠻

UEB Math/Science Code: ⠰⠰⠣⠠⠠⠁⠃⠜⠱

Ray

Symbol in Print: $\overrightarrow{AB}$

Nemeth Code: ⠁⠠⠁⠠⠃⠣⠫⠕⠻

UEB Math/Science Code: ⠰⠰⠣⠠⠠⠁⠃⠜⠘⠱

Line

Symbol in Print: $\overleftrightarrow{AB}$

Nemeth Code: ⠁⠠⠁⠠⠃⠣⠫⠪⠒⠒⠕⠻

UEB Math/Science Code: ⠰⠰⠣⠠⠠⠁⠃⠜⠨⠔⠳⠺⠗⠕

It is important when teaching students to read lines in braille that they read the entire string of characters before stating it out loud. Mathematically, teachers and students should read the above symbols aloud as "Line AB" ($\overleftrightarrow{AB}$) or "Ray AB" ($\overrightarrow{AB}$). In braille, the letters come first and the notation for what lies above the letters follows. This could cause students to read them as "AB line" or "AB ray." TSVIs should ensure math teachers are aware of this characteristic of braille notation.

Figure 5.2 An Acute Angle ($\angle ABC$) of 50.8°

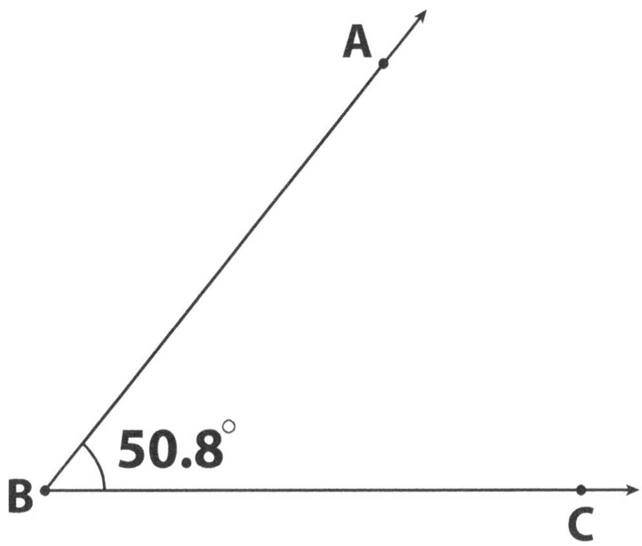

An acute angle measures less than 90º.

Angles

When three points connect (even all along the same line), it creates an *angle*. The definition of an angle is the measurement of space between two rays or lines that share a common point (referred to as the *vertex*). Angles receive their names from the three points, with the vertex as the middle point; the angle in Figure 5.2 would be $\angle ABC$. We will discuss angles and their measurement later in more detail (see also Chapter 4, "Measurement").

Angles introduce a new level of complexity to geometry, as they describe the *interaction* between two elements in space. Although students may find this difficult to fully grasp, angles make up the foundation of understanding closed shapes: The simplest possible polygon, a triangle, has three angles, three lines, and three vertices.

Congruence

The concept of *equality* in geometry is more complex than in other areas of mathematics. Most students learn about equality from an early age and associate it with numbers and measurement. *Congruence* goes beyond equality and also includes shapes and the corresponding parts of a shape. If two objects are *congruent*, they are identical, meaning they have the same size and shape even though they may be oriented differently. The symbol for congruence (≅) can be used to describe something as simple as a line segment or as complex as a dodecahedron. Congruence and equality have a mathematical relationship through various theorems. Essentially, if two objects are congruent, then the measures of the corresponding parts of the shape are equal and vice versa.

When identifying congruent figures, it is critical to identify the order of the corresponding parts. For example, in Figure 5.3, the names of the two triangles must reflect the fact that each point listed corresponds to its counterpart in the other shape: $\Delta ACB \cong \Delta FDE$. The first and last letter of the name defines the longest length in both triangles. In angles and lines, the order is less important but should be consistent throughout a problem or solution. Using the same figures, we would know that $\overline{BC} \cong \overline{DE}$ and $\angle ACB \cong \angle FDE$.

Figure 5.3 Two Congruent Triangles

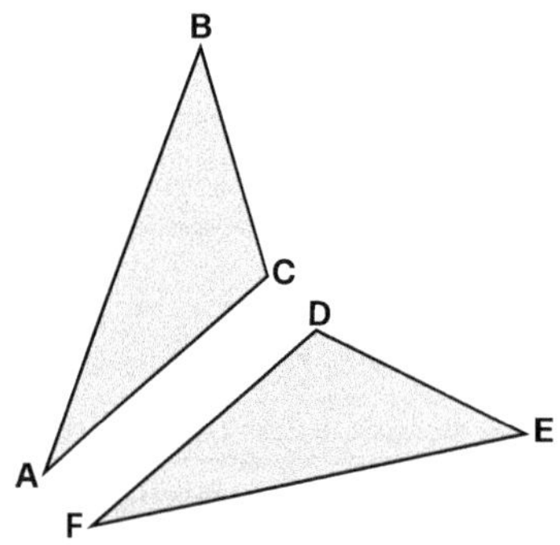

Triangles ΔACB and ΔFDE demonstrate congruence, as each angle and line segment in each shape has a corresponding angle or line segment in the other shape.

MEASUREMENT AND COMPARISON OF REAL OBJECTS

The concepts of measurement and comparison start in the earliest grades. Students learn them alongside language development when we use words such as long, short, bigger, wider, and taller to describe objects or people. Relative size is a critical concept for children with visual impairments to develop, as visual analysis may not provide a reliable comparison. Relative size is also a precursor of relative distance, which is an important part of travel using O&M skills. Without these concepts, mental mapping and keeping track of one's environment are much more challenging.

Linear Measurement: Length, Width, and Height

Linear measurement starts with identifying the length of a line segment (see Chapter 4, "Measurement"). Teachers generally introduce the concept progressively in the early grades. Students first learn to use a line with no markings to signify the distance between any two points. This practice moves into the use of a ruler to read the distance as a discrete measurement. As students learn subtraction, they make use of number lines to either count or subtract the coordinates (values) of the two points. Students build these skills slowly over time, and schools intersperse them throughout the elementary curriculum.

Completing these tasks tactually requires new tactual skills, notetaking skills, and extra practice. Whereas students with usable vision can reference locations on a line, braille readers will need the means to create embossed marks. Practice applying tactile stickers, adhesive bumps, and locking straight edges (through clamps or tape) all will be helpful for students to maintain place for measurement. Students using paper and pencil can record their marks directly on the image, while braille readers will need to record information on a braillewriter or notetaker. Note that these tasks may also be more challenging for students with low ocular motility issues or reduced horizontal visual fields, and field restriction can also be a factor in cases in which students require significant magnification. The concepts of width (horizontal) and height (vertical) are also important concepts that students will often use with little explanation by teachers needed.

The skills and activities in Sidebar 5.2 are measurement related but exist predominantly in the geometry domain of the core curriculum.

Sidebar 5.2 Strategies for Finding Distance

Distance Between Two Points

- **Prerequisite skills:** Visual/tactual tracking lines, counting, reading print/braille alphabet
- **New skills:** Tracking a line with information above or below the line, counting while also searching for a target, recording answers on the worksheet or on a separate sheet
- **Practice materials:** Draftsman Drawing Board (a product that allows students to draw on a special paper that raises with pressure), All-in-1 Board (a product that is composed of a slant board with a white dry-erase board on one side and a felt mat on the other side for materials that use Velcro; aph.org/product/all-in-one-board), Picture Maker Wheatley Tactile Diagramming Kit (an APH product that is composed of a felt board and various objects—points, lines, etc.—to place on the board; aph.org/product/picture-maker-wheatley-tactile-diagramming-kit); all are available from APH

Use of a Ruler to Measure Lines and Objects

- **Prerequisite skills:** Measuring distance between two points, counting measurements on a ruler, concept of zero or origin, notetaking skills
- **New skills:** Matching one end of the edge of the target with the zero on the ruler, method of ensuring the ruler does not move off origin (tactile or visual), method of reading ruler at target or end of edge
- **Practice materials:** Draftsman, Trifold Board (aph.org/product/trifold-board), Picture Maker Wheatley Tactile Diagramming Kit, prepared practice materials, classroom objects

Calculating Distance Through Subtraction

- **Prerequisite skills:** Measurement with counting and ruler, reading coordinates on a number line (above and below line), notetaking and recording skills

Sidebar 5.2 *(continued)*

- **Practice materials:** Tactile graphics with well-labeled grids, tactile rulers, braillewriter
- **New skills:** There should be no new tactual skills, as the activity has students apply subtraction to the concept of measurement and origin. Students should have a concept of subtraction of length to find a relative distance.

Angle Measurement

Angles can be one of the most challenging concepts to grasp in geometry. Unlike other measurements that provide a physical surface or line to measure, the defining characteristic of angles is the relationship between two lines that share a point (a vertex). The space, measured in degrees, is significantly different from measurements of distance as it is not linear but circular. The ancient Babylonians developed the concept of angle degrees by dividing the distance around a circle into 360 parts, or degrees. In higher level mathematics, students will learn to measure angles using another measurement called radians, but they will rarely complete these calculations or measurements without the aid of a calculator.

Figure 5.4 Two Types of Accessible Protractor

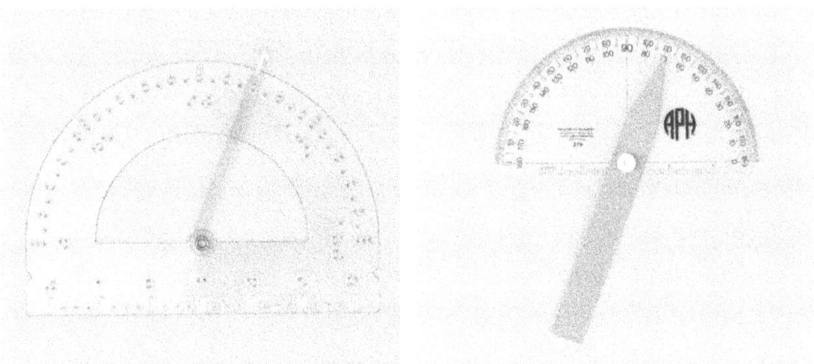

APH offers two designs of accessible protractor. The protractor on the left (aph.org/product/tactile-protractor-4-pack) has an inner wand that can be used to explore angles and features a hole at the pivot point so the origin can be marked with a pencil or pushpin. On the right is a protractor with a pivoting wand (aph.org/product/braille-large-print-protractor) that creates the corresponding angle of the degree measurement between the wand and the bottom edge.

A protractor is the tool with which users can measure angles, though students typically use protractors only for a few lessons before moving to purely arithmetic or algebraic methods of working with angles. A protractor is a semicircular shape that resembles a large letter D. To measure the angle, place the vertex of the angle at the center of the flat side (where there is normally a hole in the tool) with one side of the angle parallel to the flat side, matching with a line on the tool (see Figure 5.4). The second side of the angle will "point to" or pass through the curved side, which has the 180 degrees of the arc divided into equal spaces.

Sidebar 5.3 Steps for Drawing and an Angle With a Protractor

Steps to draw an angle with a protractor:

1. Draw a line long enough to extend the length of the protractor. Use the protractor's flat edge.
2. Match the line with the flat edge markings.
3. Mark the center (vertex) of the line.
4. With a pencil, mark the desired angle measurement along the curved edge.
5. Remove the protractor and connect the vertex to the angle mark.

Steps to measure an angle with a protractor:

1. Match one ray of the angle with the flat edge markings.
2. Line up the vertex with the center point of the flat edge.
3. Hold the protractor in position with one hand.
4. With a pencil, tactually trace around the curved edge until you reach the second ray.
5. Mark the ray with the pointer.
6. Slide your finger from the pointer to the nearest measurement.
7. Add or subtract based on the markings.

For students with blindness and visual impairments, note that many standard protractors already have raised plastic marks denoting the lines and numbers. With reasonable tactile discrimination and counting, students can use these tools effectively. Tactile protractors for angle measurement are far less accurate and precise than visual tools but are no less effective in teaching the necessary skills and concepts. That said, TSVIs and math educators should be aware of some considerations when using them. Because it is likely the tactile line is embossed, it is important not to place pressure on the tool and flatten it, which will make the line less readable. Using a protractor to draw an angle of a given size will be challenging without using a pressure-responsive paper such as that used with an APH Draftsman. The process requires several steps that embossing with manual tools makes more challenging. Without reactive paper, the process will require multiple flips of the paper and significantly more time.

Size Measurement: Area and Volume

Comparison of size in mathematics is a skill that supports estimation and approximation. The intuition developed through this skill serves as an asset to any professional using mathematics. Measurement is a core skill for comparison of size in the elementary grades and will be more important for students with visual impairments who may have difficulty using print references for comparison. The numerical measurement records and preserves what is readily visible to other students.

The language used in comparison also serves as vital vocabulary for describing the world to a child with a visual impairment. These skills start at the earliest schooling levels possible—likely even earlier for students with blindness and visual impairments. In the early grades, size comparisons will rely on physical objects and concrete examples. These examples will quickly move to worksheets, upon which students will draw targets for comparison.

These comparisons of size move from one dimension in length to two dimensions with area and finally to three dimensions with volume. Early grades employ 3-D comparison using base-10 blocks. While these manipulatives are often used to teach place value and operations, they can also be used to develop the concept of volume, particularly in basic shapes such as cubes and rectangular prisms. Teachers can have students use the unit block, rods, or flats to build cubes or rectangular prisms and then count the number of unit blocks that form the shape. This is the "square volume" of the 3-D shape and helps the students better understand the volume formulas they learn later. Another accessible tool for teaching this concept is using Stack-Ups Kit (aph.org/product/stackups-kit-spatial-reasoning-using-cubes-and-isometric-drawings) with Geometro manipulatives (aph.org/product/geometro-gs22-medium-set), both available from APH. These comparisons, as mentioned above, move from physical manipulatives to book and worksheet images of the cubes after a short time. Preparation and setup for these book or worksheet images with physical objects require extra time on the part of the TSVI and the student.

There is no accessible substitute for the physical objects in these cases, and repetition is necessary to build the required intuition and familiarity future abstract problem solving will require. The steps and process for comparison do not change for the tactile learner, but the current digital tools do not meet the unique needs of students with visual impairments because they rely heavily on dynamic visual content. Once students reach the middle grades, they will conduct more comparisons in graphs, charts, or by using numerical values (sections).

Description of Objects in Space

Relative Descriptions

The language of geometry requires the students and teacher to be very precise in describing figures or parts of figures. Casual language of normal conversation takes on the characteristics of scientific precision when used to discuss geometric objects in a plane or space. Words such as *face, side, adjacent, opposite,* and *edge* may require extra care in a math classroom context, and instructors should provide clear descriptions for the precise meaning of these terms in a mathematical context. The concepts associated with them will also need clear explanation before students encounter their use in math work. Beyond these words, common prepositions that communicate position or movement will be vital for clear understanding.

Figure 5.5 Adjacent and Opposite

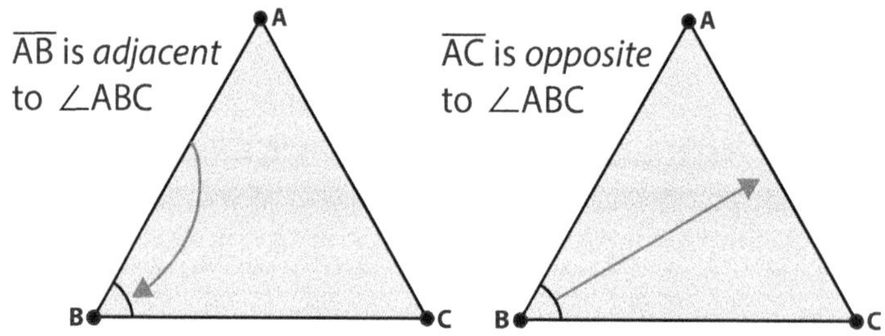

Maintaining precision of language is of vital importance in the geometry classroom.

The terms *adjacent* and *opposite* are of particular importance. In geometry, the meaning of adjacent is more specific than its typical definition of "next to" or "near." Adjacent means in contact with another geometric object. For example, in Figure 5.5, the side $\overline{AB}$ is adjacent to the angle $\angle ABC$. Opposite, likewise, has a very specific definition. An object is opposite when it is on the other side (across from the center) of the object. The side $\overline{AC}$ is opposite to $\angle ABC$.

Sidebar 5.4 Common Terminology to Describe Objects in Space

- *Side*: A line segment that connects two vertices of a 2-D geometric shape. (i.e., the base/bottom of a triangle, or the third base line of the square the makes a baseball diamond)
- *Face*: One flat surface of a 3-D geometric shape that connects to the other flat surfaces that make up the shape (i.e., the base of a pyramid, or the with the number 6 of a common die cube)
- *Edge*: The line segment that joins two faces of a 3-D geometric shape
- *Adjacent*: A line, angle, or point that is directly next to and in contact with the referenced object
- *Opposite*: A line, angle, or point that is across a shape or open space from the referenced object

Absolute Coordinate Systems

The Cartesian coordinate system is the most commonly used absolute coordinate graphing system in school mathematics. It uses the x-axis and y-axis to describe objects in a 2-D space. Including a z-axis can help represent three dimensions. Mathematicians write the points in a 2-D Cartesian coordinate as (x, y) and call the point located at (0, 0) the *origin*.

The conceptual understanding required to use or make a graph is a fundamental part of the Common Core State Standards. Students with visual impairments will already be familiar with the concepts behind absolute coordinate systems, as their O&M specialists frequently teach them map grids. Coordinate systems are powerful tools for math students who are blind or visually impaired. A coordinate system can provide many types of information:

- *Relative location*: (5, 1) is in the first quadrant and next to the y-axis.
- *Spatial understanding*: (3, −4) is to the right and down from the origin.

- *Distance*: $(7, 5)$ and $(-3, 6)$ are at least 10 units away from each other.

Cartesian coordinates allow students to compute important values in algebra, such as slope and intercept. In geometry, coordinates are used for a variety of situations such as drawing polygons using precise points, finding the length and midpoint of a line segment, and aiding in transformations within the Cartesian plane.

Figure 5.6 Points and a Line Segment on a Cartesian Coordinate Plane

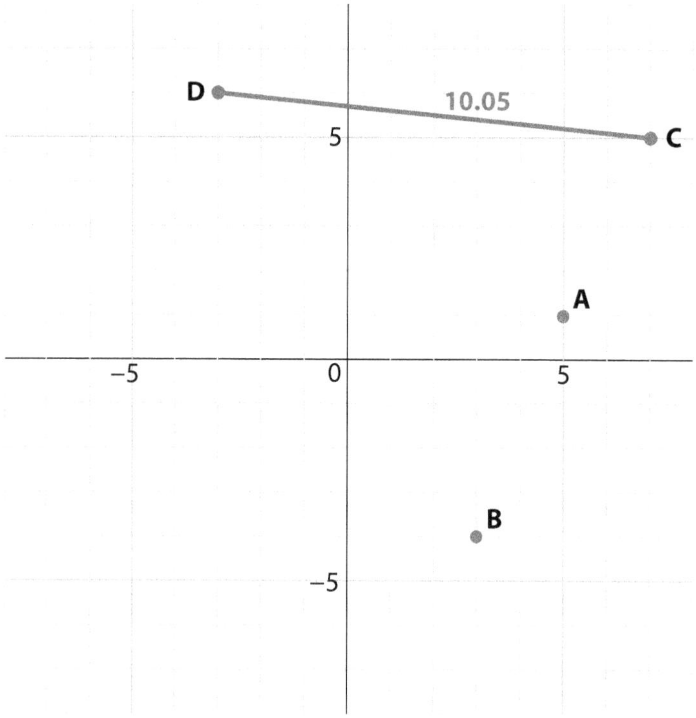

Points plotted in the first second and fourth quadrant of a Cartesian coordinate plane. The distance of a line can be calculated based on the coordinates and the distance formula.

2-D Shapes

Shapes are a foundational concept in geometry. Identifying special shapes, their specific characteristics, and whether multiple shapes have similar characteristics are all the result of discrete measurement. Geometers (individuals skilled in the use of geometry) were historically interested in identifying "perfect" versions of shapes. These perfect shapes are what we normally associate with the typical triangle, square, or pentagon. In the perfect shape, all sides and all angles have the same measurement. Today, we describe these perfect shapes as *regular*. In a regular shape, the sides are *equilateral* (all sides are congruent) and the angles are *equiangular* (all angles are congruent). Any shape that does not have all equal sides and angles is *irregular*. There is one special irregular shape related to triangles and trapezoids, *isosceles*, which means having exactly two equal sides. Shapes with no equal sides or angles are *scalene*, and classrooms tend to explore this concept using triangles.

Preschool toys and activities introduce these shapes to children at the earliest ages. The shapes return year after year in the curriculum as concepts and as placeholding symbols. For a TSVI, it is useful to understand these shapes in terms of their geometric relevance. Their mathematical use, properties, and manipulation all fall within the domain of the classroom math teacher.

A core principle of teaching geometry is that one cannot assume any size or measurement from the diagram, so using tactile shapes should not pose a problem for students and TSVIs if the shapes have changeable labels for the points, angles, and sides. Triangles and their measurements will play a particularly major role in the conceptual understanding of trigonometry, so students should receive instruction that prepares them to interact with tactile diagrams and problems that explore the properties of this shape.

Triangles

Triangles are the simplest 2-D shapes in geometry. As such, triangles are also the shape students will study the most. In geometry, one can simplify almost all other polygons down to a combination of triangles. The triangles that students see in the early grades tend to be equilateral, having all sides of equal length. Students will then encounter isosceles triangles, which have two sides that are equal in length. Finally, students will engage scalene triangles which have no sides equal, and start to apply the fact that the interior angles must add up to 180 degrees. Students will also classify triangles by their angles: *acute* (having all angles less than 90°), *right* (having one angle that is exactly 90°), and *obtuse* (having an angle that is larger than 90°). Both scalene and isosceles triangles can potentially be acute, right, or obtuse, but all equilateral triangles must be acute.

Figure 5.7 Three Types of Triangles

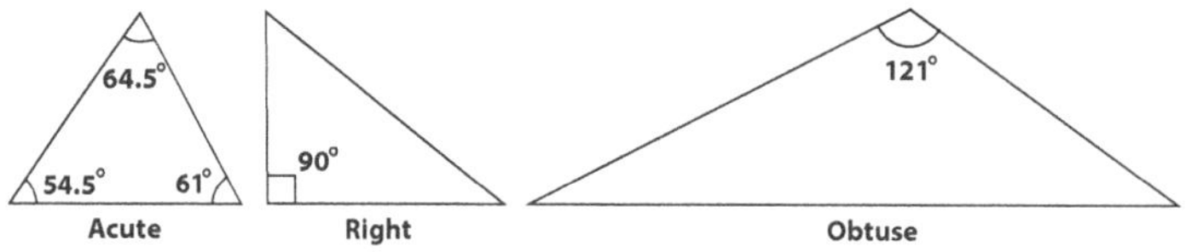

Acute, right, and obtuse triangles are defined by their angles.

Quadrilaterals

Quadrilaterals are four (quad)-sided (lateral) shapes. The addition of the extra point significantly increases the variety of classifiable quadrilaterals. The special classifications of quadrilaterals are the kite, trapezoid, parallelogram, rhombus, rectangle, square, and isosceles trapezoid.

Sidebar 5.5 The Quadrilateral Family

While all *quadrilaterals* have four sides, they can be subdivided based upon their attributes.

- *Parallelograms* are quadrilaterals whose opposite sides are parallel and congruent and opposite angles are congruent. A parallelogram's diagonals bisect each other.
    - *Rectangles* are special types of parallelograms where all four angles are right angles, and the diagonals are congruent.
    - *Rhombuses* are special types of parallelograms that have four congruent sides, the diagonals are perpendicular to each other, and the diagonals bisect opposite angles.
    - *Squares* are parallelograms that have all the attributes of both rectangles and rhombuses.
- *Trapezoids* have one pair of opposite sides that are parallel.
    - An *isosceles trapezoid* exists when the pair of opposite side legs that are not parallel are congruent.
- *Kites* have two pairs of adjacent congruent sides, and its opposite sides are neither congruent nor parallel.

Mathematicians simply refer to all other four-sided shapes as quadrilaterals. When studying quadrilaterals, students will measure them, compare sides for congruence, compare angles for congruence, check for parallel sides, and check for perpendicular sides. Students can conduct these evaluations through proofs, with paper and compass, or by using digital geometry tools (such as the Desmos Geometry Tool, found at desmos.com/geometry). All of these methods come with unique challenges, but digital tools are not inherently more beneficial than simpler traditional tools. For students using tactile materials, pressure-reactive paper and a compass will work well. For students with low vision, screen magnification works well with digital tools, and desktop magnification can support traditional methods.

Figure 5.8 Quadrilaterals From Reflected Triangles

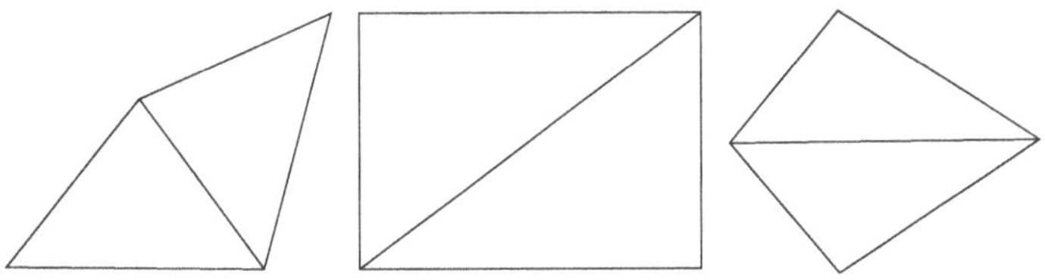

Two reflected triangles can be combined to create a quadrilateral

Polygons With Five or More Sides

Students do not explore polygons with five or more sides to the same extent as they do triangles and quadrilaterals, and teachers seldom emphasize their classification in the classroom beyond those figures that are regular (having congruent sides and interior angles) and irregular (everything else). Students will interact with these shapes when working in-class examples and homework problems, but that is normally the limit in K–12 education. Students should know how to calculate angle measurements (interior and exterior), number of sides, perimeter, and sometimes area. The greatest challenge with these calculations will be the need to isolate portions of the shape into recognizable special triangles or quadrilaterals. (For more discussion of the part–whole relationship between advanced shapes and simpler shapes, see Chapter 3, "Expansion of Numbers and Operations.")

Circles

The circle is a shape of unique form and interest in geometry. It is the logical extension of polygons in that it has a perfectly smooth shape and as a result has no "sides." One can define the circle in that all points are the same distance from some center point. Mathematicians always name circles in geometry by the shape and the center point (e.g., "Circle B," which has B as the center point).

When a diagram combines circles with triangles (or other polygons), it increases the complexity of resultant proofs and calculations. Lines classify circles by the number of times they intersect. Adding line segments to a circle will introduce *tangent* and *secant* lines: Tangent lines intersect with a circle at a single point on its arc, and secant lines intersect twice, crossing through the interior of the circle. These lines are related to trigonometry functions (discussed later in this chapter), but teachers normally introduce them long before trigonometric relationships.

For students who need them, producing tactile versions of diagrams that involve circles can easily become crowded with information, making them potentially difficult to decipher. In order to support student learning, the TSVI may need to create multiple versions of the circle with the parts of the embedded information removed. For example, if a circle has multiple lines drawn within the circle (chords, diameters, radii), then it may be appropriate to create simplified tactile graphics that include only a few of the parts. The TSVI still needs to have the original, complex version available so that their students can compare the graphics to make decisions.

Students start interacting with circles in the early grades by finding equal portions and thinking of them as pie charts. They will relate to these circles as objects such as wheels or hula hoops. In the middle grades, students learn about radius and diameter. They will explore these two measurements and eventually find the circumference (perimeter) and area. Students who are blind or visually impaired will be able to understand circles by making them. In the late elementary and middle school grades, students will be introduced to ways of drawing a circle, such as using a pin with string or with a specialized tool, the compass. Essentially, whether a student uses a pin with string or a compass, the same process is used to draw the circle: The student selects a point that serves as the stationary point while a specific length (the radius) is determined, and the circle is formed by tracing around the center point while maintaining the same distance from the circle. While students with low vision may be able to use typical compasses, students without functional vision will need to use specialized tools such as APH's Tactile Compass for Math and Art (aph.org/product/tactile-compass-for-math-and-art). Coupled with a tactile board such as APH's Draftsman or TactileDoodle (aph.org/product/tactile-doodle-kit), the students can create their own circles. The APH Hive site has a free Tactile Aids for Mathematics course (aphhive.org/#/courses/enroll/7; login required) that demonstrates this and other common tools, as well as tutorial videos that can provide users with clear demonstrations.

In middle and high school, students will demonstrate the ability to find arc length and section areas by using an angle measurement, relating back to the elementary activities of cutting a piece of pie. Advanced math students will explore the lengths of *chords* (interior lines terminating on the circumference) based on their distance from the center. Teachers present most of the academic work related to circles as a carefully drawn diagram or through calculations.

MEASUREMENT AND COMPARISON OF MATHEMATICAL REPRESENTATIONS

Measurement is a critical aspect of the geometry curriculum. While the concepts of measurement are covered in depth in Chapter 4 ("Measurement"), a discussion of the applications of measurement within geometry is warranted in this chapter. For the purposes of this chapter, we will focus upon measurements that are taught by the development of formulas based upon specific types of shapes.

Perimeter and Area

Finding the perimeter or area of an object, done with the simplest tools, involves measuring each side of the object and using those measurements to determine the outcome. This activity demonstrates one of the fundamental practical applications of geometry: A farmer trying to calculate the amount of seed he needs (area) or the amount of fencing to surround his property (perimeter) is putting the principles of geometry to work. Chapter 4 provides details on how students with blindness or visual impairments can use measurement to find perimeter and area.

In math classrooms, direct measurement instruction falls away in the early grades. Soon, teachers quickly replace this with graphical representation that includes numbers to label the length of sides. At the middle and high school level, these numbers become letters, introducing algebraic computation.

In complex drawings and figures, such as the one in Sidebar 5.6, the content can quickly become crowded, which poses a challenge for students with blindness and visual impairments. The most effective method to decrease this complexity for such students using print materials is to redraw the figure in parts, allowing the students to solve problems as they go. When TSVIs prepare these materials, they should be careful not to fundamentally alter the intended problem-solving procedure and definitely should consult with the content teacher when considering any changes.

Sidebar 5.6 Decomposing a Complex Geometric Problem

Problem: Find the area and perimeter of the geometric shape:

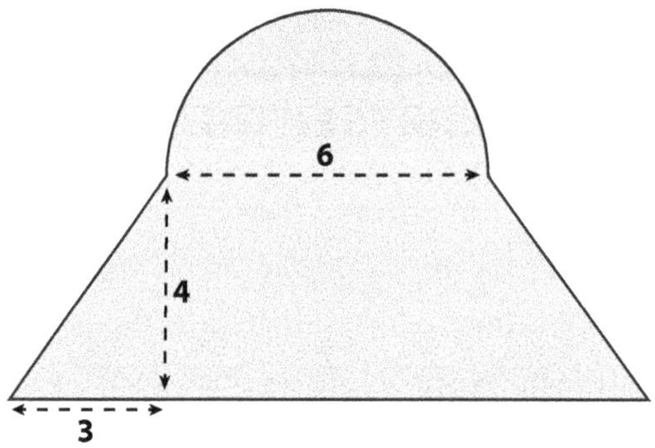

One can think of the above figure as a combination of as few as two and as many as four geometric shapes. The smallest, most efficient solution requires students to see a semicircle on top of an isosceles trapezoid. One can also see it as a rectangle, semicircle, and two triangles. Regardless of how they divided the figure, students would be familiar with the area and perimeter functions for all of these shapes from prior exercises.

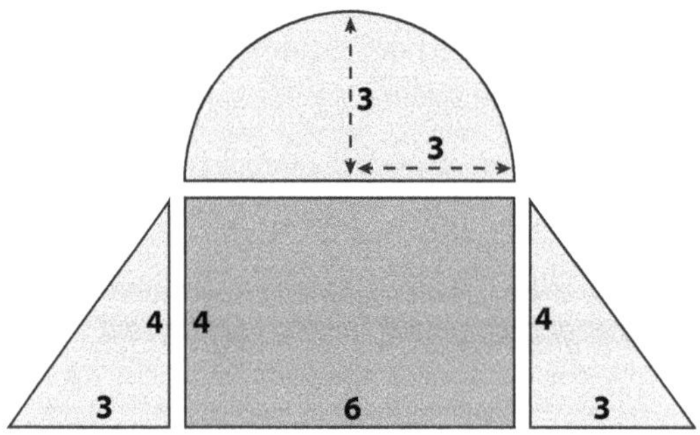

A standard process for solving this problem would involve the students redrawing the shape with paper and pencil, effectively "taking apart" the

Sidebar 5.6 *(continued)*

shape. To prepare an equivalent set of tools for a tactile learner, it is important to create each piece separately with given measurements. To adapt this problem, the teacher would need to create tactile versions of the original graphic and then the semicircle, rectangle, isosceles trapezoid, and two triangles separately (allowing the student the choice in whether they want to use the trapezoid or deconstruct it further into a rectangle and two triangles).

The teacher could tape together the original graphic's individual pieces, allowing the students to fold the object where different shapes occur. It is possible to similarly tape Tactile Tangrams (available from APH at aph.org/product/tactile-tangrams-kit), allowing more space for folding. (To do this, allow a small gap between the pieces before applying the tape, or use flexible fabric tape.) It is possible to use any prefabricated shape if the sizing is appropriate for the linked edges and the inclusion of braille labels.

When students receive these problems using digital graphics only, the interpretation of the figure and the redrawing process are not accessible to students using tactile materials. It is possible, but at the sacrifice of considerable effort and time. The TSVI would need to develop a strategy along with the math teacher to provide complex geometric figures using accessible materials. This may include using paper as described in this example, creating tactile graphics, or using the Geometro tool as appropriate.

Volume

Volume takes the two dimensions of an area calculation and adds a third. The underlying mathematics do not add significant complexity: The calculation for volume involves simple multiplication. The challenge for students with visual impairments lies in the diagrams and figures. Representing a 3-D shape in a 2-D tactile graphic creates the possibility that a student will misapprehend what the figures represent. As such, we recommend students use physical models whenever possible. Reusing the same models will help the student develop a strong familiarity with the process.

However, one must observe caution when using perfect or regular shapes as models that might create misconceptions for the student. For example, if a student tries to measure a rectangular solid with Sides 2, 3, and 4 and only has a cube available to demonstrate the problem, they may develop misconceptions regarding the properties of cubes or rectangular solids. The students using print materials see objects with varied sides, though they should not assume anything from pictures; ideally, the same would be true for students working with tactile models.

Algebraic Equations for Geometric Principles

Students' perception of geometry as a subject changes drastically when teachers introduce algebra to describe the measurements, areas, and changes in shapes. Students will learn to find points, midpoints, and slopes using nothing but numbers and variables. They can complete the same process looking at the objects drawn on a grid or through measurement, but the algebraic approach requires more ability on the part of the student to comprehend abstraction. The algebraic emphasis in K–12 math means that students will have to master geometric and algebraic approaches and understand the connections between them.

Midpoint and Bisection

One can find a midpoint through various processes, including physically, algebraically, and geometrically. The most basic approach to finding the midpoint is to use the midpoint theorem and equation based on two (x, y) coordinates:

$$\left(\frac{x_1+x_2}{2}, \frac{y_1+y_2}{2}\right)$$

The algebraic formula describes the process of counting the number of units of length on the x- and y-axis from the first to the second point and dividing that by 2. When completing this process using algebra, there are few barriers to access. When finding a midpoint on a line segment in one dimension, such as a number line, the equation omits the y-axis:

$$\frac{x_1+x_2}{2}$$

Teachers can adapt these equations to a horizontal orientation to better align with how they would appear in UEB Math/Science or Nemeth code (see Figure 5.9). To find the midpoint of the line from 3 to 11, start with the midpoint formula written all on a single line (as it would be in braille).

Figure 5.9 Adapting Equations to Horizontal Orientation to Align With Braille Codes

Horizontal Orientation:

$$M = \frac{(x_1+x_2)}{2}$$

$$M = \frac{3+11}{2}$$

$$M = \frac{14}{2}$$

$$M = 7$$

Nemeth Code:

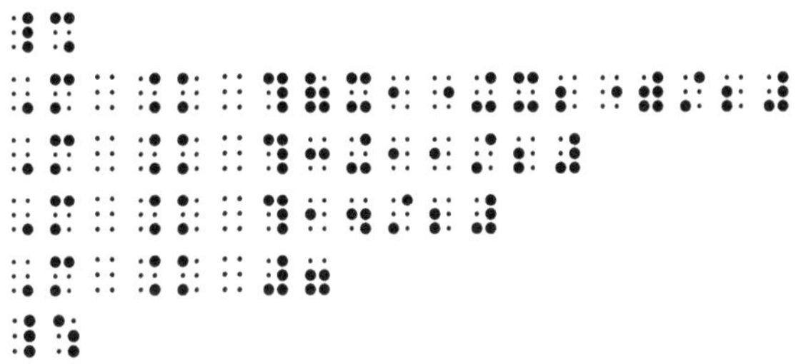

UEB Math/Science:

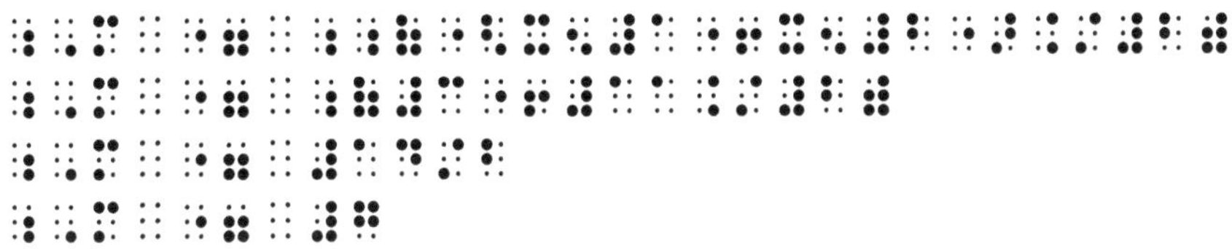

In looking at Figure 5.9, this is essentially half of the midpoint formula; thus, if a student can complete this for the x-values, they should be able to complete it for the y-values. When completing this same process in two dimensions, the core equation appears more complex when viewed in coordinate form . Solving this problem in "pieces" keeps the added complexity manageable. This management style of complexity is useful in all higher level math and is even more useful when solving equations in braille. It is important to understand, though, that the approach takes practice and some level of cognitive load or organized methodology to keep the pieces together. In all math, good methods work all the time, and the same piecemeal approach works for three dimensions, such as with .

Slope

The slope of a line, taught in algebra, is a core component of the slope–intercept equation of a line. (See Chapter 6, "Foundations of Algebra," for a more in-depth examination.) The topic shows up in coordinate geometry and is frequently includes in topics involving parallel and perpendicular lines. These topics will almost always include a grid with geometric shapes graphed on them (most commonly triangles, quadrilaterals, and circles). Students will learn to find a slope and/or check whether two slopes are equal (parallel) or reciprocal (perpendicular).

HIGHER ORDER GEOMETRY CONCEPTS

Trigonometry

Trigonometry is the study of the measurements of trigons (triangles). When students begin to study the relationship between the sizes of angles and sides in a triangle, they will encounter several significant changes in how they understand geometry and represent it with mathematical notation. Trigonometry first appears as a topic in geometry class, but it is also a significant part of precalculus and calculus. The ability to evaluate and manipulate trigonometric functions, identities, and laws will be particularly critical skills for students who intend to enter STEM-related fields.

The three basic trigonometric functions (sine, cosine, and tangent) may be the first time students will see a multiletter function in mathematics. Second-year algebra introduces the logarithm (log) and precalculus frequently introduces limits (lim), which are a core of calculus. Students will need to read and recognize the three letters and combine them with a variety of math symbols (numbers, Greek letters, function notation, etc.). This will become intuitive with practice, but teachers need to introduce it in advance. The relationships of the angles and trigonometric identities are related to the sides of a right triangle that are adjacent or opposite the given angle (though not all trigonometry concerns right triangles).

Students can find the values (coordinates) of the trigonometric functions using a unit circle. The unit circle is a circle of radius 1 drawn around the origin that includes common angles in degrees, radians, and sine and cosine values. In most cases, students memorize the values present in the circle. However, when computing the inverse of trigonometric functions, they may need to rotate the circle clockwise or counterclockwise to identify the correct value. A unit circle labeled with all necessary values for reference can be incredibly crowded and difficult to create or read in braille. In these cases, a table of values is an easier solution. Table 5.1 contains the same information as a unit circle in a simpler form.

Table 5.1 Sine and Cosine Values

Angle (Degrees)	Angle (Radians)	Sine	Cosine
0°	0	0	1
30°	$\frac{\pi}{6}$	$\frac{1}{2}$	$\frac{\sqrt{3}}{2}$
45°	$\frac{\pi}{4}$	$\frac{\sqrt{2}}{2}$	$\frac{\sqrt{2}}{2}$
60°	$\frac{\pi}{3}$	$\frac{\sqrt{3}}{2}$	$\frac{1}{2}$
90°	$\frac{\pi}{2}$	1	0
120°	$\frac{2\pi}{3}$	$\frac{\sqrt{3}}{2}$	$-\frac{1}{2}$
135°	$\frac{3\pi}{4}$	$\frac{\sqrt{2}}{2}$	$-\frac{\sqrt{2}}{2}$
150°	$\frac{5\pi}{6}$	$\frac{1}{2}$	$-\frac{\sqrt{3}}{2}$
180°	π	0	−1

The writing of trigonometric functions and the solving of equations involving them will often require careful use of parentheses. The placement of exponents, new uses of pi, and the addition of theta as a variable will add new challenges to the brailled materials, so teachers should prepare students who use braille for the additional complexity. Math classes traditionally move slowly and deliberately through these materials, but some integrated math courses may move in and out of the topic more quickly.

Advanced Geometric Shapes in 2-D and 3-D

While most of geometry focuses on studying special figures, some problems in area and volume will use more complex shapes. One will almost always be able to divide these shapes into simpler shapes. The diagrams for these shapes will often use dotted lines to demonstrate lines hidden by the foreground of the shape.

There is no clear way to represent these diagrams in 2-D tactile diagrams. Any effort will cause portions of the object to be obscured, which is typically represented by dotted lines in print. Different lines can be used to replicate the visual information, but the result is a 2-D picture that requires far more effort to understand than the actual object would for the tactile learner. The best way to construct these shapes in 3-D is by using the principles of *nets*. Nets are 2-D shapes that can be cut and folded to form a 3-D object. When carefully planned, the labels and braille can be added before assembly or after, provided sufficient space is available. Students can also use a tool such as the Geometro to create flat nets that can then be used to create the 3-D shape (see Chapter 4, "Measurement"). Teachers should take great care when making these 3-D models, including defining a clear top and bottom and always applying braille using this up/down orientation. Avoid applying braille on angles or along lines that would require the reader to rotate the tactile graphic. Maintaining relative position and orientation takes additional working memory and could make the problem more difficult for the student. Remaking diagrams larger using labels with the same orientation can also be a valuable adjustment for students who have low vision.

3-D Shapes

When studying geometry, students briefly study geometric objects in three dimensions. Most work with 3-D objects involves calculating surface area (the area of all external faces) or volume (as discussed earlier). These shapes will most often involve expansions on four-sided shapes and circles into the third dimension. Some shapes will not simply extend into the third dimension, but all points will converge to a single point (seen with a pyramid or cone). These extended and converging figures will often include interior sides or diagonal lines that assist with computation. Often, these interior lines will represent an extra layer of complexity, and students will use them to find the height of the figure. In these cases, the 3-D tactile objects require an open face or hinge to view the inside. A 2-D cross section of the 3-D figure can convey the same information. Cross sections are an important topic in algebra class as well.

Tools are available for the construction of nets and masks for 3-D objects, such as Geometro and Magna-Tiles (magnatiles.com). Unfortunately, as these tools tend to focus on regular polygons, it is difficult to diagram book problems without some (if not significant) modification. This modification will include braille labels for the vertices, braille labels for the edges or measurements, and the addition of interior components (such as altitude lines from the base to the top at a right angle). Advanced models are available for 3-D printing at sources such as Tactiles (tactiles.eu) or through construction using Tinkercad (tinkercad.com).

Regular Polyhedra

Like the regular shapes of two dimensions, regular polyhedra are three-dimensional shapes that have all faces congruent (instead of equal sides in 2-D). Students may be familiar with the nontraditional dice with more than six sides that are common in modern tabletop roleplaying games. While all of these are polyhedra, there are only five *regular* polyhedra, called the platonic solids. These shapes can be great practice for students to explore, build, and identify. Triangles comprise three of the polyhedra (4-sided tetrahedron, 8-sided octahedron, 20-sided icosahedron), squares comprise one (6-sided cube), and pentagons comprise one (12-sided dodecahedron). If students have the language and tools to explore these objects from an early age, they will be familiar concepts when the students encounter them in geometry class. Parents can introduce the activity of building the shapes with tools such as Geometro from APH or Magna-Tiles to their children from as early as age 2. These tools also allow children to build with a reduced fear of knocking things over or apart. The repeated construction of these shapes can be a familiar puzzle akin to solving a Rubik's Cube. The spatial understanding that comes with the activity will also create a stronger understanding of three-dimensional objects in space.

Figure 5.10 3-D Rendered Object in Desmos Software

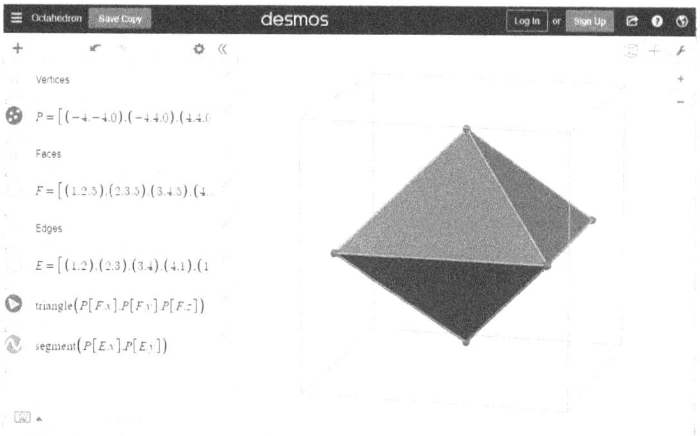

Desmos can mathematically illustrate 3-D shapes, such as this rectangular octahedron. *Image was created with the graphing calculator, used with permission from Desmos Studio PBC*

TRANSFORMATION

Both algebra and geometry classes have topics related to changing (transforming) shapes. Most textbooks present these transformations on a coordinate x–y plane. This coordinate system allows students to discuss points in terms of their coordinates instead of just the label (name) of the point. When discussing transformations, terms include *reflect* (flip), *translate* (slide), *rotate* (turn), and *dilate* (scale). Each of these transformations changes the location (flip/reflect, turn/rotate, slide) or the size of the original shape. As the names of these transformations suggest, students can easily explore them with manipulatives, with the exception of scaling.

Reflect (Flip)

The process of "flipping" (reflecting) a shape involves picking it up with all but one side remaining on the surface and laying it down with that side on the opposite face of the shape. A simple demonstration could be done with any card (business, credit, playing). First, lay the card on the desk. Then, lift the card, keeping one edge still in contact with the desk. Finally, flip over the card so that its opposite side faces up. Completing this process without a card face staying in contact with the surface is called *reflecting*, and the reflecting line (the edge that stays in contact with the surface) defines where to place the reflected shape: The shape remains at the same distance and retains the same alignment to the reflecting line as when it started on the opposite side.

Flipping shapes using tactile materials is easiest with a cutout shape. Teachers can demonstrate this process using a shape with one edge taped to the surface. Students can complete their own reflections easily if they receive multiple copies of the same shape. APH's Graphic Aid for Mathematics (aph.org/product/graphic-aid-for-mathematics-2) can further aid this process. As students complete more complex reflections, the reflecting line will not be part of the shape; it may be the x- or y-axis or an arbitrary line defined by the problem. The Graphic Aid will allow students to easily diagram and explore this so they can pay careful attention to distances and position.

Translate (Slide)

Students can think of translating a shape as sliding it in a direction on its coordinate plane. In early academic tasks, there will be a steady progression from these basics to diagonal sliding. Students can first slide the shape in the horizontal direction and then in the vertical direction to result in a diagonal slide. In more advanced mathematics, students will receive an angle and a distance. This angle and distance measurement is the basis of polar coordinates used in subjects such as trigonometry and force diagrams in physics. This form of sliding requires a strong understanding of angles and experience with turning, discussed below.

Teachers can easily demonstrate sliding with a cutout shape and an x–y axis. Using the x–y axis is critical because sliding has no real meaning without some point of reference. O&M lessons with furniture can help build this concept in the early grades as students learn to slide objects toward another object, such as a chair under a table. Reference points such as the door or another piece of furniture can serve as the target. Instead of measuring distance in traditional units, students can use steps or tiles as a unit of measurement. For a fun twist involving measurement units based on body parts, the teacher can discuss hands (4 inches, or roughly 10 cm) and feet (the more familiar 12 inches, equivalent to 30.48 cm).

Rotate (Turn)

Turning, or rotating, a shape involves moving the object around a single point. That point is normally one of the vertices of the shape. It is easy to replicate this process by pinning a shape to a board by the rotation point. A student then moves it clockwise or counterclockwise a percentage of a full rotation or by a number of degrees. Because angle measurements are the only factor, students must be familiar with different ways to measure angles. Students can practice concepts such as quarter turn and half turn with O&M lessons and facing directions. Clock directions and cardinal directions can also be useful to help students understand the principles.

The idea of turning is relatively simple. When used with coordinate geometry—as it almost always is eventually—the calculations can be a bit more complex. It is essential that students always check their work with tactile diagrams or a graph of their starting and finishing shape. As with learning reflections, cutout shapes should support these activities.

Dilation (Scale)

Scaling, also known as dilation, is quite different from the three previous transformations. In those transformations, the object retains its shape and size. Scaling involves reducing or enlarging the object in size. Normally, a person can do this with a reference point (similar to turning), but a modifier ($\times 2$ or $\frac{1}{2}$) increases the distance along the horizontal or vertical axis. Replicating this transformation with physical materials is much more challenging. The best way to create a physical demonstration would be to use a geoboard or pegboard, laying out the original shape and then the resulting shape. This would allow students to compare the two and see the point of reference as well. Scaling will be much easier for students with low vision to grasp, as modern classroom tools allow teachers to fluidly change shapes. The skills used in scaling are important in a number of careers ranging from graphic arts to engineering.

CONCLUSION

As geometry is the study of physical space and shapes, teachers can theoretically construct and explore everything in a way that is accessible to all students. The modern digital tools and 3-D images for geometry are where the greatest challenges lie, as they are commonplace and many are not accessible for students with visual impairments. Fortunately, the majority of geometry content focuses on 2-D shapes, which can be re-created easily. The creation of multiple graphics can be time consuming, but they are the single most important factor in helping students develop the intuition and understanding of geometry. Research is exploring and has shown some measurable benefits of digital interactive tools to improve spatial reasoning skills among students in different groups (Nurlaelah, 2018; Sezen Yüksel, 2017; Nagy-Kondor, 2017). With new products improving refreshable tactile graphics, it is possible that our students will have access to dynamic 2-D geometry tools in the near future. While these tools aid concept development and exploration, proof of knowledge in testing and assessment will still rely heavily on static graphics.

Because we live in a world constructed of geometric shapes, developing an intuition and understanding of geometry will reward visually impaired students for the rest of their lives. Students may find the subject overwhelmingly challenging due to its visual nature, but the language and concepts they encounter early on through O&M prepare them well. At an early age, students first encounter tactile literacy around geometry in braille books even before formal literacy; likewise, if a child experiences a steady stream of tactile graphics that assist with understanding the material, they will have access to advanced subjects, making way for careers in engineering, design, architecture, and the hard sciences—all of which build and rely on the principles of geometry at some level. The more traditional proof-oriented geometry taught prior to 2000 added the benefits of formal logic and reasoning to the topic. However, although the subject emphasizes algebraic equations in the contemporary curricula, the benefits of geometry far outweigh the efforts required to make it available to students with visual impairments.

REFERENCES

Heiberg, J. L. (1956). *The thirteen books of Euclid's elements* (T. L. Heath, Trans.). Dover Publications. (Original work published 1908.)

Hilbert, D. (1950). *The foundations of geometry* (E. J. Townsend, Trans.). Open Court Publishing Company. (Original work published 1902.)

Nagy-Kondor, R. (2017). Spatial ability: Measurement and development. In M. S. Khine (Ed.), *Visual-spatial ability in STEM education: Transforming research into practice* (pp. 35–58). Springer Nature. https://doi.org/10.1007/978-3-319-44385-0_3

National Council of Teachers of Mathematics. (2000). *Principles and standards for school mathematics*.

Nurlaelah, E. (2018). Improving junior high school students' spatial reasoning ability through model eliciting activities with Cabri 3D. *International Education Studies, 11*(1), 148–154.

Sezen Yüksel, N. (2017). Measuring spatial visualization: Test development study. In M. S. Khine (Ed.), *Visual-spatial ability in STEM education: Transforming research into practice* (pp. 59–84). Springer Nature.

CHAPTER 6

FOUNDATIONS OF ALGEBRA

Derrick W. Smith

For many adults, *algebra* can be an emotionally loaded term. It evokes visceral fear from memories of a mathematics class where numbers and letters were mixed up in some type of strange arrangement that they did not understand. Algebra is the first step students take into more abstract thinking about mathematics, and the additional unknowns that come from working with variables such as x and y can be intimidating. In truth, even *defining* algebra is quite difficult, as it is a complex branch of mathematics that is constantly evolving.

The elementary algebra taught in K–12 education is only the beginning of what students can accomplish with algebraic reasoning (Kaput, Carraher, & Blanton, 2008). It is the foundation for geometry, statistics, calculus, and all higher mathematics. The learning of algebra is so important that students who plan to continue in postsecondary education, particularly in STEM-related fields, are required to take at least two years of algebra study. Understanding algebra and developing algebraic reasoning is often considered the "gateway" to many professional careers and is becoming increasingly necessary for access to and success in all types of careers, even those outside of STEM fields (National Council for Teachers of Mathematics [NCTM], 2018, p. 10). Therefore, all students, including those with visual impairments, should receive a solid foundation in algebra as part of the K–12 educational curricula.

WHAT IS ALGEBRA?

What seems like a simple question is quite complex and has been written about extensively. The NCTM defines algebra as "the systematic exploration of the arithmetic of expressions involving unknown or variable quantities as well as methods for finding those unknowns or determining relationships between or among variables" (2018, p. 45). They go on to explain that algebra is "a systematic study of patterns" and primarily focuses on functions (p. 45). Functions are an algebraic concept that explains the relationship between one quantity and another. They can be categorized as an algebraic expression and are often expressed graphically.

For the teacher of students with visual impairments (TSVI), even these definitions may not provide a clear view of the actual definition of algebra. Ultimately, algebra is the mathematical science used to solve complex problems in which the goal determining an unknown.

Contrary to popular belief, we use algebraic reasoning in our lives all the time. Each of the following problems uses a principle of algebra:

- How does the distance that a person walked depend on the time walked? [Answer: $d = rt$]
- What time will we need to leave our house in order to get to work on time?
- What is the maximum number of tickets we need to sell to recoup the costs of an event to earn a profit?
- How much will I earn this week if I worked 40 hours at normal pay and 8 hours of overtime? [Answer: $P = (40 \times \$10.00) + (8 \times (\$10.00 \times 1.5))$]

In this way, algebra is not only the foundation of more advanced mathematics; it is also the foundation of our understanding of the world around us.

WHEN SHOULD ALGEBRA INSTRUCTION BEGIN?

When most people think of algebra, they think of the mathematics that started between seventh and ninth grade, either in the form of a prealgebra or an Algebra 1 course. Research has long held that it is around this development timeframe when students are able to shift from arithmetic (which is more concrete) to algebra (which is more abstract) (Warren, Trigueros, & Ursini, 2016). However, there has been a stronger focus on developing "algebraic thinking" (or algebraic reasoning) as early as kindergarten while students are learning arithmetic. The NCTM *Principles and Standards of School Mathematics* (2000) has long advocated for the inclusion of algebra across the grades instead of a few isolated courses on the topic. The *Common Core State*

Standards for Mathematics (National Governors Association Center for Best Practices & Council of Chief State School Officers, 2010) changed the landscape of mathematics education across the United States; and it is notable that the authors include the need for "operations and algebraic thinking" as early as kindergarten; we continue to see these topics flowing through all the elementary standards through fifth grade. Beginning in sixth grade, "prealgebra" concepts of "expressions and equations" are introduced and expanded on from that point forward. Thus, the answer to the question is deceptively straightforward: Algebra—or at least algebraic reasoning—begins as soon as mathematics instruction does (Van de Walle, Karp, & Bay-Williams, 2019).

For the TSVI, this means that they will need to work closely with the mathematics teacher (elementary and secondary) to ensure that their students with visual impairments have access to the algebraic reasoning concepts. Many early algebraic reasoning activities require students to work with tokens, stacking cubes, base-10 blocks, or hand drawing to show their reasoning when solving problems. Many of these manipulatives are innately accessible because they are tangible objects. Drawing is accessible with the use of appropriate tools. For example, students can use a Perkins Brailler to create number lines and tally marks or the Draftsman Tactile Drawing Board from the American Printing House for the Blind, or APH (aph.org/product/draftsman-tactile-drawing-board) coupled with drawing tools when appropriate (ruler, compass, etc.) to "sketch" their mathematical ideas. In other cases, alternatives will need to be provided for elementary students to be able to complete these very important learning activities.

MAJOR CONSIDERATIONS FOR TEACHING ALGEBRA TO STUDENTS WITH VISUAL IMPAIRMENTS

As with any subarea of mathematics, algebra has its own distinct considerations for students with visual impairments. Algebra and algebraic thinking can be quite challenging for many students. as it uses the foundations of arithmetic to explore generalization and symbolization (Kaput, Carraher, & Blanton, 2008). This is challenging because it often moves beyond the rote memorization of number facts and explores the structures of the number system based upon arithmetic. Combine this with a focus on patterns, relations, functions, and mathematical modeling, and algebra can be quite daunting for all students.

This presents four major challenges particular to new algebra learners. Starting when prealgebra formally begins, typically in the middle school grades, there is a major shift from concrete concepts and arithmetic to more abstract ideas. This creates a challenge for all students, but students with visual impairments may have a more difficult time because the new focus can compound challenges initially encountered during earlier mathematics content. Any limitation or misapprehension a student experiences in elementary mathematics tends to create major challenges for future learning as mathematics becomes more complex.

A second consideration for students with visual impairments is that students must quickly learn new symbols and how to use them as they progress through the algebra curriculum. This can create unique barriers that teachers should be prepared to help these students overcome. As a basic example, the use of superscripts (exponents) and subscripts, root symbols (with indices), matrices, logarithms, etc. all require accommodations for print size. Even symbols such as equals (=) and approximately (≈) are incredibly similar and will require scrutiny for students to differentiate between them. For students who read braille, there will be an increase in new braille symbols that will need to be learned and, more importantly, understood. For instance, students who use braille must understand the purposes of special indicators used in effects, such as with complex fractions (fractions that include operations and/or variables) and radicals. Even the use of superscripts and subscripts will have braille rules students have not encountered due to the complexity of written algebra.

A third major challenge faced in algebra is the use of graphs as a fundamental concept. Graphing functions (equations and inequalities) is a concept seen and utilized throughout the algebra curriculum. Obviously, the ability to draw graphs can be a major challenge for students with any type of visual impairment; an ancillary issue connected to graphs is that many classes depend upon graphing calculators to complete many of the graphing tasks. There are accessible graphing calculators, including the Orion TI-84 Plus Talking Graphing Calculator (orbitresearch.com/product/orionti-84-plus) and the Desmos Graphing Calculator (desmos.com/calculator) that provide audio and/or vibration feedback and can also connect to some braille embossers to print out the graphs as tactile graphics. However, these tools require specialized instruction by the TSVI to ensure the students have access to all the available functions.

Another challenge is that many algebraic concepts rely upon spatial arrangements, particularly in how they are taught. For example, students may learn to use the First, Outer, Inner, Last (FOIL) method for multiplying two binomials, and they also may be taught to draw lines as they use the distributive property to solve the problem.

For students with significant vision loss, however, this may not be an option, as writing braille and drawing at the same time would be quite challenging. Another example is the study of matrices, which is primarily a way to solve systems of equations using spatial arrangements. Adapting material for these arrangements and for rational expressions (fractional expressions and equations), and various other concepts is addressed later in this chapter.

Despite these four significant challenges and others, students with visual impairments can learn and become quite proficient at algebra with careful use of appropriate accommodations. As with all mathematics instruction, the general education math teacher and the TSVI must collaborate to ensure that the students have access to equitable learning opportunities.

STRATEGIES FOR TEACHING MAJOR ALGEBRA CONCEPTS

Students' understanding of certain major algebra concepts may be especially impacted by their blindness or visual impairments. The following topics within this chapter are not exhaustive but should provide an instructor with basic strategies for accommodating these algebra students.

Spatial Arrangements

One of the first accommodations that must be addressed is how to solve problems that use spatial arrangements. There are a host of algebraic topics that rely upon some type of spatial arrangement, such as equations, systems of linear equations, synthetic division, rational expressions (fractions with variables), and matrices. For example, students are often taught a specific way to solve single-variable equations that uses spatial arrangements (see Figure 6.1, top). Using this procedure, students are required to align parts of the equation vertically so that it functions like a type of basic subtraction problem. It is possible to align a problem this way in braille, but it requires careful planning as students who use braille work the problem. To help address this, there are multiple ways to solve equations that do not require spatial arrangements. The second arrangement in Figure 6.1 demonstrates how to solve the same problem while keeping all information on one horizontal line, which better supports students who write braille. TSVIs and math instructors will need to collaborate to avoid spatial arrangements for solving problems as much as possible.

Figure 6.1 Vertical and Horizontal Arrangements With UEB Braille Equivalents

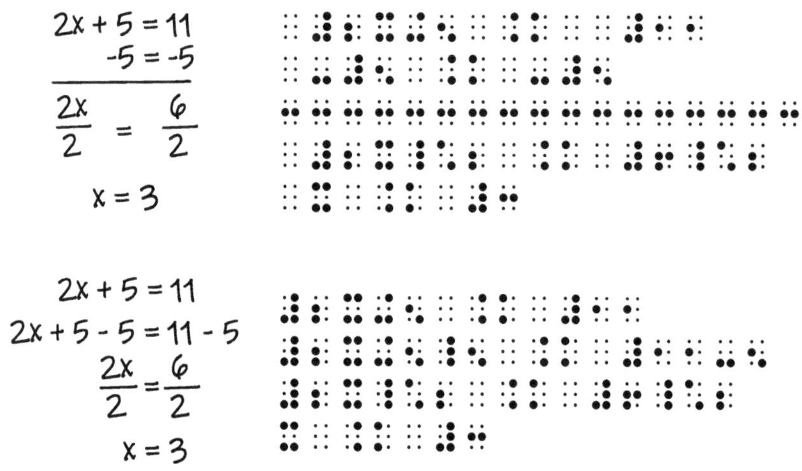

Arranging an equation vertically (top) can create additional work for students using braille. Arranging horizontally (bottom) keeps each step of the formula on one line, which is much easier to render in braille.

Another example of spatial arrangements appears in the study of matrices and determinates, typically found in the Algebra 2 curriculum. At first glance, a matrix provides an obvious challenge, as it requires a careful spatial arrangement (see Figure 6.2).

Figure 6.2 A Simple 3×3 Matrix

$$\begin{bmatrix} 3 & -2 & 4 \\ 8 & 11 & 13 \\ 6 & -2 & 7 \end{bmatrix}$$

Matrices are almost always presented as a spatial arrangement.

While alignment in print of matrices is simple, the arrangement in braille can be quite cumbersome, particularly when negative symbols, fractions, or decimals are involved. Braille has specific rules that require that the matrices' indicators are aligned. As the numbers within the matrix may include negative symbols, decimals, or have a different number of digits, the alignment may require extra preplanning by the TSVI and the students, as alignment is important when working with matrices (particularly arithmetic between matrices), and the extra braille cells required for these math terms adds an extra challenge.

Fortunately, clear solutions to these challenges exist. First, matrices and determinants *can* be written using either Nemeth code or the Unified English Braille (UEB) Math/Science code. For example, in Figure 6.3, we have written the matrices from Figure 6.2 in both Nemeth code and UEB Math/Science. However, students, in collaboration with their TSVI and math teacher, may create a simplified, alternative method for writing matrices that is less formal (see Figure 6.3). For example, a student may not include the symbols for the matrices' grouping symbol (Student Alternative Version 1) or may simply use the parenthesis symbol (Student Alternative Version 2).

TSVIs not only will need to provide extra direct instruction to introduce the new math syntax for matrices, but they will also need to focus on the preplanning required to write these correctly, as they require special spacing. However, as a secondary accommodation, a TSVI may consider providing the students with a tangible way to write numerals in a spatial arrangement using an accessibility tool. Examples would include devices such as the Brannan Cubarithm Slate (aph.org/product/brannan-cubarithm-slate) or other tools that allow for arrangement of numbers such as the Math Window (mathwindow.com). With the ongoing development of new multiline braille devices, students will be able to more easily write math problems that require special spatial arrangements.

Figure 6.3 A Simple 3×3 Matrix Rendered in Braille

Nemeth Code Example

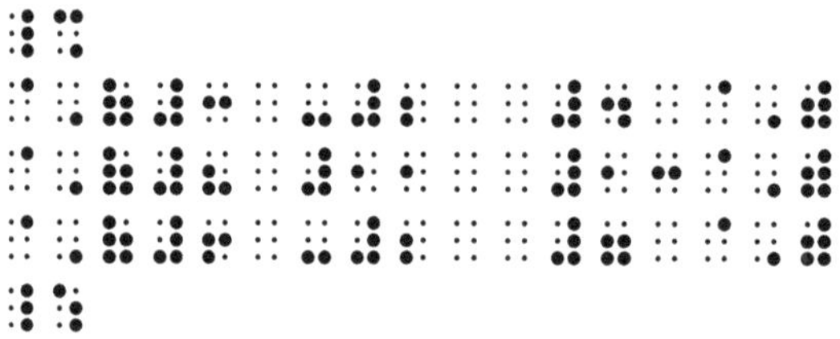

UEB Math/Science Example

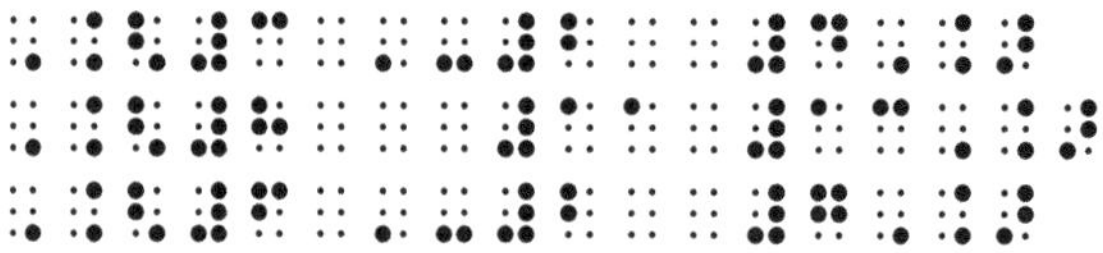

Student Alternative Version 1

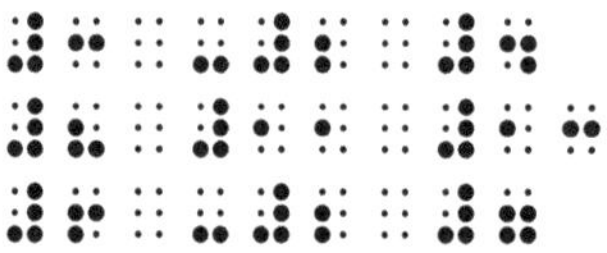

Student Alternative Version 2

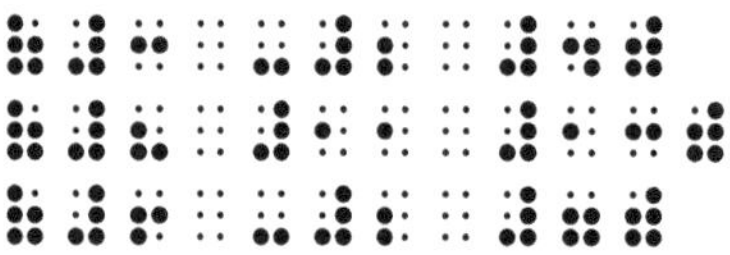

Students may want to use their own way of aligning a matrix, and teachers should be aware of this while ensuring that student adaptations do not introduce misunderstandings or miscalculations.

Spatially Focused Instructional Strategies

Within mathematics education, educators sometimes devise instructional strategies to help students remember a specific process. One example is the mnemonic device of "PEMDAS" or "Please Excuse My Dear Aunt Sally," which illustrates the order of operations used to solve expressions: Parentheses, Exponents, Multiplication, Division, Addition, and Subtraction. With this example, the only tangible accommodation for students with visual impairments is that their teachers must instruct them to read the entire expression carefully before they start solving the problems. Further, they must be aware of grouping symbols (such as parentheses) and exponents before they start working with other operations.

Another algebra topic that tends to use some type of instructional strategy is when multiplying two binomials, such as $(x + 4)(2x - 8)$. There are a variety of strategies that educators employ to help students learn this process. One prominent example is using the aforementioned FOIL method. The FOIL method is a simple way to help students remember to solve the problem using the distributive property. For students with visual impairments, the math teacher and/or TSVI will need to make sure they understand these positional terms, which may be vague with just an oral explanation. (Often, teachers will draw lines matching the first term of the first binomial to the first term of the second binomial. Typically, this would be accessible for students with low vision as long as the drawings do not become too cluttered.) There are also other methods where teachers use a "box method" that involves the students creating a 2 × 2 grid, placing the terms in the box, and computing the parts of the problem; however, this would not be an accessible strategy for students with visual impairments.

Figure 6.4 The FOIL Method for Solving the Binomial $(x + 5)(2x - 3)$

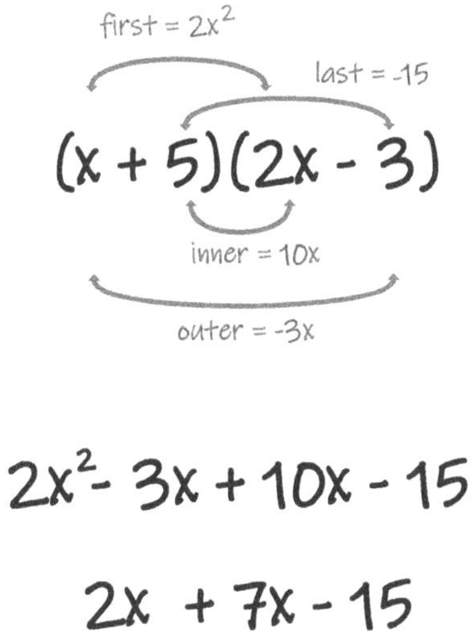

Students are often taught to use lines to keep track of binomial multiplication using the FOIL method, which can be accessible if the drawings are not too cluttered.

Figure 6.5 The Box Method for Solving the Binomial $(x + 5)(2x - 3)$

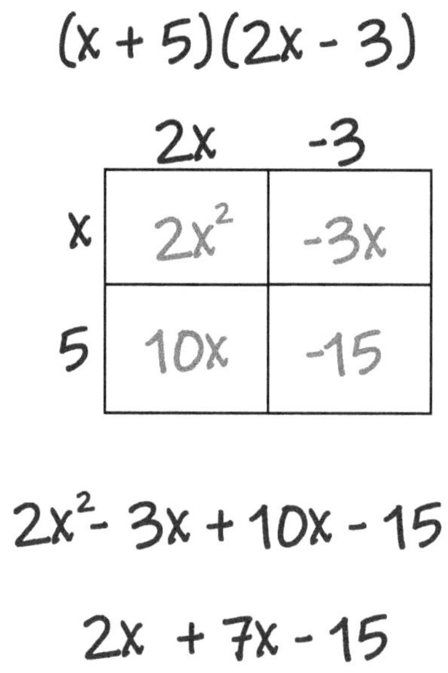

Due to its visual nature, the box method can present significant challenges for students with visual impairments.

While these methods may be helpful to some students, some students will prefer solving the problem using the actual mathematical property at hand: the distributive property. It may be easier for students with visual impairments to learn that they will take the first term in the first binomial and multiply it by *each* term in the second binomial. Then, they will take the second term in the first binomial and multiply it by each term in the second binomial. This is essentially what the FOIL method does, but instead of a visual strategy, it relies on memorization of the purely mathematical approach.

Across the mathematics curriculum, there is a large array of other approaches that have been devised to help students solve problems that use unique spatial constructs which will have limited or no use for students with visual impairments. For example, to write superscripts (i.e., exponents) and subscripts, many students rely on a spatial construct in which a term is "shifted" up or down in relation to another term. This is also a concern when working with rational numbers and/or expressions (fractions). As fractions are written spatially with a "top" (numerator) and "bottom" (denominator), there are many strategies that depend on spatial aspects when solving. There are even some algebra concepts that use strategies to solve problems that are similar to long division (i.e., long division and synthetic division of expressions). For these strategies, spatial alignment of numbers is often crucial to the strategy. Therefore, the TSVI and the math teacher should collaborate with the students to ensure they have an efficient strategy.

While all of these spatially focused processes can be written in braille, the TSVI will need to collaborate with the mathematics teacher to determine the best strategy for the students with visual impairments. For example, the TSVI and the students may need to develop a writing strategy that is simpler than the formal way to write the problems according to the braille code rules. The TSVI may need to explain to the mathematics teacher to help them understand the challenges that students may have with spatial arrangements. Even with something as simple as a rational expression, the braille is written horizontally, so the students may not clearly understand the "top and bottom" concept of fractions clearly. Collaboration, flexibility, and responsiveness to the individual student's needs by both the TSVI and mathematics teacher are critically important for many of these spatially focused strategies.

Another common spatially focused instructional strategy found in algebra is the use manipulatives that represent value. *Algebra tiles* have different sizes that represent numerical values, while *algebra balances* use weight to represent numerical value that can be balanced using a scale. Both approaches use tangible objects that are often differentiated by color, size, and/or text on the objects. These concepts are a vision-based instructional strategy primarily used for factoring and solving equations. Often, teachers will use the manipulatives while introducing these concepts and may then switch to a digital version of the manipulatives later. For students with visual impairments, the tangible tools may be accessible with some minor modifications. For the "algebra balances," the manipulatives need to be able to be differentiated based on texture or size. Braille labels could be added to the objects, but this may interfere with the weights of the objects. (The APH Tactile Algebra Tiles product, available at aph.org/product/tactile-algebra-tile, is already accessible; see Figure 6.6.) As items in this manipulative rely upon size and shape, they can be made accessible for most students.

Figure 6.6 Tactile Tiles for Algebra

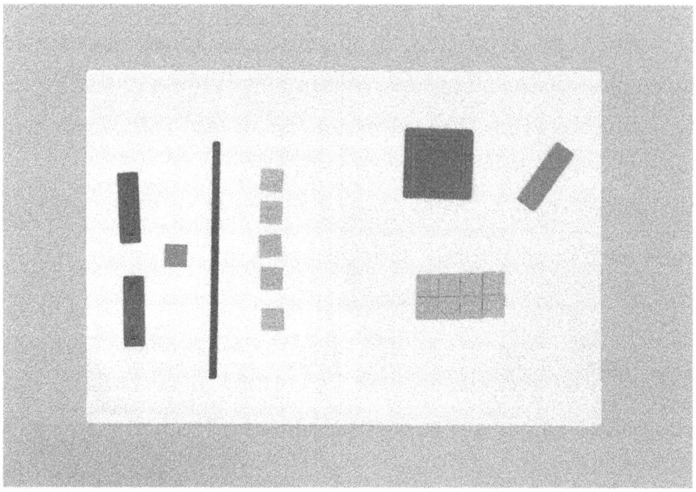

Algebra tiles, such as these from APH, are a common manipulative that students use to understand algebra concepts.

Graphing

Graphing is a prominent aspect of the algebra curriculum that students must master to be successful. Graphs are a visual way to show all possible solutions for a function (or equation) with more than one variable. Most of the graphs in K–12 education focus on two dimensions (typically x and y) and are graphed on a Cartesian coordinate plane formed by two perpendicular lines. In advanced algebra classes, students will begin to explore graphing in three-dimensional spaces (typically x, y, and z planes). Instruction with graphing typically begins with students being required to graph points on the coordinate plane and then graph the equations themselves on grid. Once mastered, the curriculum transitions to using graphing calculators to graph more advanced functions.

For students with low vision, graphing any type of line may have some specific challenges. First, most grid paper utilizes very light or thin grid lines that may be in blue or green. For some students with low vision, using commercially available grid paper may not be accessible. Therefore, the TSVI may need to provide the students with dark-lined grid paper that uses thicker, blue or black lines for the grids. (The thickness and darkness of the grid lines should be determined based upon the student's individual needs.) The students may also need to use a pen or marker that can make darker and thicker points and lines, and some students may benefit from the use of a different color (such as blue) to help with contrast.

For blind students, the ability to "draw" graphs can be challenging, but with the right types of accommodations, it is possible. In general, these students will need to either use tactile grid paper or some other type of tactile grid that they can place points and the graphs upon. The next sections provide specific strategies for graphing various algebra concepts.

Figure 6.7 A Graph of the Function $y = x + 4$

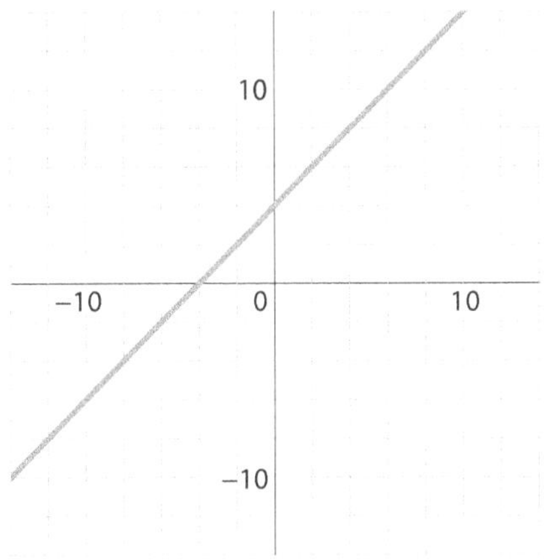

In graphing a function, addition and subtraction will move the line's position in relation to the origin, while multiplication and division change the slope.

Strategies for Graphing Lines

When graphing linear equations or functions, the primary output is a straight line (see Figure 6.7 for the graph of $y = x + 4$). The mathematics teacher will most likely teach the students multiple strategies for graphing equations such as having the students use the equation to create a table, finding the x- and y-intercepts, or using the y-intercept and the slope of the equation. All of these strategies are based on the equation, and thus the students will need to learn the algebra processes to use the equations to be able to plot points. The only major accommodation needed for students who are blind or visually impaired may be in helping the students devise a system to create an x–y table that they can then implement individually when they need it.

Students with low vision should avail themselves of the accommodations noted above. To draw the line, the TSVI will need to ensure that the students can effectively plot points on the coordinate plane and then use a ruler to construct the line that goes through the distinct points. The students should also learn to label the points used on the graph to ensure that the teacher can find those points when accessing the graph.

For blind students, there are a few ways to graph an equation. The first way is to use a fixed graph tool and have the students graph the line using pushpins or thumbtacks, rubber bands, and/or strings. An example of such a tool is APH's Graphic Aid for Mathematics (aph.org/product/graphic-aid-for-mathematics-2; see Chapter 4, "Measurement"), which is a rubber grid that allows students to attach pushpins or thumbtacks to fixed points on the grid. There will need to be some accommodations when using this tool (or other similar tools such as geoboards). For example, having the students create the x-axis and y-axis may be appropriate when introducing the topic, but over time this activity will no longer be necessary; instead, the students will need to determine the *origin* (the intersection point of the two axes).

When plotting the graph, it is recommended that the students use a flat thumbtack to denote the origin and then use pushpins to represent the other points on the graph. This will help both the students and the teacher to differentiate between the origin and points on the graphed line. Once the students place points on the graph (a minimum of two for straight lines, more for curved graphs), they can then use string, rubber bands, or something else to form the line on the graph. In order to be able to turn in the graph to the teacher, the TSVI may need to teach the students how to use a camera (or the camera on a phone or tablet) to take a picture of each graph before moving on to the next one. The inability to submit the actual graphs is the primary inhibitor of using board tools (which are more cumbersome and static than paper versions), but it is excellent for in-class instruction and practice.

Another strategy that allows students to submit their graphs is to provide them with tactile grid paper (such as APH's Embossed Grid Paper, aph.org/search-results/?fwp_search_term=embossed+grid+paper). For this strategy, the TSVI will need to supply the students with a rubber or corkboard square pad that is the same size as the tactile grid paper. Here are the steps for graphing using tactile grid paper:

1. The students would place the tactile grid paper on the pad and determine where to place the origin with a flat thumbtack.
2. The students should then use a pencil or pen to circle the origin. This will mark the origin so the teacher will be able to determine where it is later.

3. Then, using pushpins, the students will select multiple points that lie on the function's graph.
4. At this point, there are two options for submitting the assignment:
   a. The students can use APH Graph Benders (aph.org/product/graph-benders), string, or something similar to create the graph and then take a picture of it. Then, they can submit the digital photograph to the teacher.
   b. An alternative option is to use a slate and stylus to mark what problem this graph represents (which can be done beforehand by the teacher). The student can simply submit the tactile grid paper to the math teacher. The teacher will be able to determine where the origin is, as the student will circle it. Then they can look to find the holes created by the pushpins (which may require looking at the paper held up to the light). This should allow the teacher to determine what points the student created for the graph so they can assess the assignment. (Ultimately, teachers depend upon the graphed points more than the actual graph to determine whether the graph is correct.)

Obviously, using tactile grid paper requires providing the students with enough sheets to complete their assignments. There is also a limitation based upon the number of grid lines on the page. Therefore, the TSVI and the mathematics teacher will need to teach the students to think about the placement of the original thumbtack as it may not always need to be in the middle of the page. There will also be situations for which each grid line will need to represent a different scale to accommodate the equation (such as each line representing 2 or 5).

There are a few notes that must be addressed here: There are some graphs that use straight lines but form other shapes, such as a "V" shape of equations that involve absolute value or a parabola (U-shape) for quadratic equations. In graphing these types of graphs, students may need to use many pushpins to form the graph. While students need experience creating these graphs, they also need experience reading professionally created graphs as tactile graphics. While many textbooks will provide these types of examples, the TSVI may need to work with the mathematics teacher to create such examples to make these graphics accessible.

Strategies for Graphing Inequalities

After learning how to graph equations, students will naturally shift to graphing inequalities. The procedures for graphing inequalities use the basic strategies for graphing an equality but include additional components to represent the inequality. For example, if an inequality is "greater than" (>) or "less than" (<), then the answers to the inequality do *not* include the line as part of the answer. In print, this is denoted by drawing a dashed line. The solid line denotes inequalities that are "greater than and equal to" (≥) or "lesser than and equal to" (≤). The second unique aspect when graphing inequalities is that a section of the area must be "shaded" to represent the area that includes all possible solutions. As an example, Figure 6.8 provides the graph of $y > 2x + 3$ and shows the broken line and the shaded area to the left.

Figure 6.8 A Graph of the Inequality $y > 2x + 3$

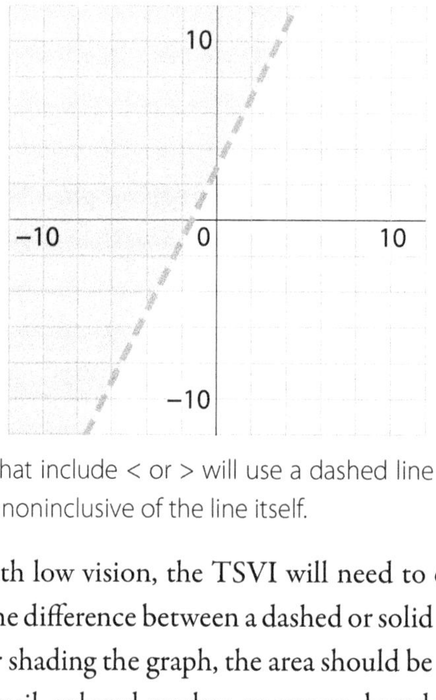

Inequalities that include < or > will use a dashed line to represent that they are noninclusive of the line itself.

For students with low vision, the TSVI will need to ensure that the students can perceive the difference between a dashed or solid line based upon the inequality signs. For shading the graph, the area should be shaded using either a pencil, colored pencil, colored marker, or crayon, based upon their individual preferences. For blind students, if they are using a graphing tool, they will need to determine a way to create a dashed line versus a solid graph. For example, they may use a rubber band for solid lines and a string for a dashed line; then, they must denote the shaded area. (While there are many ways to do this, the simplest way to do this would be just to put some type of special pin on the side of the line that denotes the shading.) If the students are using tactile grid paper, students could use the same strategy, or they could arbitrarily "shade" the area using a pencil. They also could place a ruler on the line and then mark the region to be shaded with perpendicular lines. In the end, they merely need a means to denote which side of the graph should be shaded. When graphing a single inequality, this would be the best strategy.

Strategies for Graphing Systems of Equations and Inequalities

Up to this point, we have been discussing graphing strategies for a single graph, yet there are many situations in which multiple graphs occur in combination. Students typically find this when studying systems, which requires determining any shared solutions between two or more functions. When solving a system of two linear graphs, the intersection of the two lines (if it exists) is the primary objective of the system. Blind students will need to graph one line and create another line with string or a similar medium. Then, they will need to graph the second line and attempt to determine where the two lines intersect. Once the students have the coordinates, they will need to check that their solution works for both functions.

Note: As a side note, systems of equations can also be solved algebraically (without graphs), so a student who is blind may benefit more from solving the systems without graphing, depending on the goal of the activity. If the goal is to understand the relationship between the equations and their graphical counterpart, then the students need to be able to graph the lines. However, if the students have a choice, they may choose to solve the problem algebraically.

Another common activity is to solve systems of inequalities by graphing the lines, shading the appropriate areas, and then determining which shaded areas overlap. For blind students, they will follow the same procedures noted above, with the objective of determining what shaded areas overlap and/or are shared. This will require them to place all the lines on the graph first and then determine the shared shaded area. Again, students can solve this algebraically, which may be a better option for students who are blind.

Further, it should be noted that systems of equations and inequalities can involve more than just two functions and will require careful planning during instruction and assignments using the procedures noted above.

Strategies for Graphing Curves

During their first algebra course, students will be introduced to graphing quadratic functions (functions with an exponent of 2, such as $f(x) = x^2 - 5x + 6$) that will result in a curved shape called a *parabola* (see Figure 6.9). As students progress through the algebra curriculum, they will learn about a variety of graphs that include different shapes of curves; the basic principles for drawing these graphs are similar across the different types. To understand this, students first should receive high-quality tactile graphic exemplars of the graphs so they can learn the basic "shapes" of graphs based upon the type of function.

Figure 6.9 A Graph of the Parabola for the Function $f(x) = x^2 - 5x + 6$

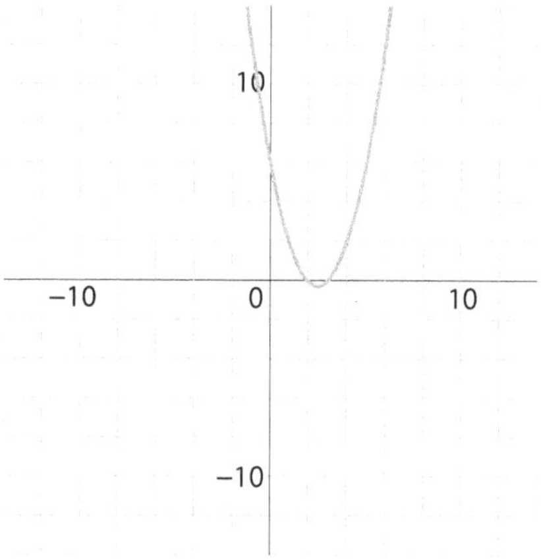

Graphing quadratic functions is a key part of formal algebra instruction.

When tasked with creating the graphs, students with low vision and blind students will need to clearly understand a few major principles.

1. Students will need to know the basic shape of the function, such as quadratic functions, that form a parabola.
2. Students will need to understand how to use the scale of the grid as some advanced graphs will use fractional coordinates that will need to be approximated on the graph.
3. Students will need to be able to create an appropriate x–y table that has enough points to "see" the curve. There will be a need to use many more points (pushpins for blind students) to create this curve.

Graphing Calculators

As noted earlier, as students progress through the mathematics curriculum, they become less dependent on drawing graphs by hand and switch to using graphing calculator tools. Graphing calculators can not only create the graphs but also provide students with different forms of data (tables that include points on the graph, zooming features, tracing features, etc.). Several options for these tools exist for students who have low vision or are blind.

For students with low vision who cannot use graphing calculators, there is a vast array of graphing calculators available for use on tablets or computers and can allow the students to access the information while using their device's accessibility tools (such as zooming).

For blind students, there are currently fewer options for graphing calculators that are accessible. The most notable are the Orion TI-84 Plus Talking Graphing Calculator from Orbit Research and the Desmos Graphing Calculator.

- The Orion TI-84 Plus is built upon a standard TI-84 but includes multiple accessibility features such as speech output, audio and haptic feedback, and even the ability to emboss graphs on a compatible braille embosser.
- The Desmos Graphing Calculator access is available either through their website (desmos.com) or by downloading their accompanying app onto tablets or smartphones. Both are accessible to screen readers and embossing graphics. This calculator is also accessible to electronic braille displays in either Nemeth code or UEB Math/Science. While this graphing calculator provides many features, it does require access through a computing device and training on the special commands to make it accessible.

Spoken Mathematics

Algebra concepts use a combination of mathematical syntax that students learn at the same time they are exposed to arithmetic involving new symbols and constructs. Within many secondary algebra classrooms, mathematics teachers rely heavily upon writing out problems on the board while teaching the concepts. Symbols such as operators, the negative sign, grouping symbols (such as parentheses, brackets, braces, absolute value, etc.), rational numbers (fractions), exponents, and radicals are just a sampling of the unique syntax that students will need to be able to write as they take notes within the classroom.

For students with low vision, the ability to efficiently look from the board to their paper as needed to take notes may be challenging. Even with distance magnifiers, the constant need to switch viewing from their paper to the board—regardless of the magnification technology used—can complicate their ability to keep pace with their peers. Blind students will need to rely solely on their hearing as they take notes.

A central consideration to the instruction of more advanced mathematics is that there are many situations in which the way that a teacher may say the math expression may not convey what was written on the board. Take this example:

$$\frac{3}{2x-5}$$

A teacher may write this on the board and say something like, "3 over 2x minus 5." Based on that description, a student who is blind could easily misinterpret the rational expression:

$$\frac{3}{2x} - 5$$

Because of imprecise language, the student is not able to concretely conclude what the teacher means. In another example using the radical symbol, the teacher may read it aloud as "the square root of 4x minus 7." Without visual information, there's no way for a student who is blind to determine which of these two is meant:

$$\sqrt{4x-7}$$

or

$$\sqrt{4x} - 7$$

Again, the student would need to guess at the teacher's meaning, and while these expressions look similar, they have completely different mathematical meanings.

For people who know any math braille code, these problems explain the reason that both UEB Math/Science and Nemeth code use opening and terminating indicators. For example, in UEB Math/Science, we would write

$$\sqrt{4x} - 7$$

as

⠰⠩⠼⠙⠭⠬⠐⠤⠼⠛

In Nemeth code, we would write it as

⠸⠩⠀⠜⠲⠭⠻⠤⠶⠀⠸⠱

In the Nemeth code example above, ⠜ represents the beginning of the radical and ⠻ indicates that the radical has terminated (with ⠲⠭ representing $4x$). The use of opening and closing symbols in Nemeth code clarifies any confusion that students might encounter through spoken mathematics.

Because mathematics teachers are not typically versed in braille codes, the TSVI will need to provide some instruction to help the students with visual impairments be able to access the information on the board. First, the TSVI must explain to the mathematics teacher the challenges the students may have in being able to keep up with the written work from the board. The combination of challenges in quickly seeing the information on the board and then quickly switching to the paper (for students with low vision) or not being able to see the board at all (blind students) will require explanation. The TSVI will need to explain that braille takes longer to write, as it has many required features that have no equivalent print symbols. However, explaining the use and purposes of opening and closing (terminating) symbols used in braille should make sense to the mathematics teacher.

Additionally, the TSVI should explain to the teacher that math braille codes provide a syntax that even the regular mathematics teacher can use to help make the information written on the board clear to not only the blind students, but possibly to all students. There are at least two formal systems of "spoken mathematics": Dr. Nemeth's Mathspeak (Nemeth, A., 2013) and Dr. Chang's Spoken Mathematics ("Larry's Speakeasy") (Chang, 1983). Both systems have their strengths and weaknesses, but both are essentially a means to speak about mathematics that allows students to write math correctly without looking at the board. Sidebar 6.1 contains some examples that the TSVI could share with the mathematics teacher if Nemeth code is being used, along with an explanation as to how these help all students, in particular braille readers. If the student is using UEB Math/Science, the TSVI would need to create a similar document that uses the UEB code indicators. The individual mathematics teacher can decide how to make this system more efficient for their students.

Sidebar 6.1 Spoken Math Examples That Demonstrate How Clarity Can Be Achieved

$(a+) - c$	"Open parenthesis, a plus b, close parenthesis, minus c" or "Open paren, a plus b, close, minus c."
$-(3 - 6) - 2$	"Negative, open parenthesis (or paren), 3 minus 6, close parenthesis, minus 2."
$\|-12\|$	"The absolute value of negative 12" or "Open absolute value, negative 12, close."
$\|3x + 7\| + 2$	"The absolute value of 3x plus 7, close absolute value, plus 2" or "Open absolute value, 3x plus 7, close, plus 2."
$\{x + y\}$	"Open bracket, x plus y, close bracket."
$[3x + 2y]$	"Open brace, 3x plus 2y, close brace."
$\sqrt{2x - 3}$	"The square root of 2x minus 3" or "Open square root, 2x minus 3, close root."
$\sqrt{2x} - 3$	"Open square root, 2x, close root, minus 3."
$\sqrt[3]{-9} + 5$	"Open cube root, −9, close, plus 5."
$\frac{x+y}{z}$	"Open fraction, x plus y, over (or divided by) z, close fraction."
$\frac{2x+1}{3} - 4$	"Open fraction, 2x plus 1 over 3, close fraction, minus 4."
$x^{2y+3} - 5$	"x to the 2y plus 3, close exponent, minus 5."
3 ± 8	"3 plus/minus 8."
$x_y - 3$	"x sub y, close sub, minus 3."
$\frac{-b \pm \sqrt{b^2 - 4ac}}{2a}$	"Open fraction, negative b plus/minus, square root of b squared minus 4ac, close root, over 2a, close fraction."

CONCLUSION

Algebra is attainable for all students, including those with visual impairments. While this chapter is not exhaustive of the challenges these students may face while learning algebraic concepts, it is a list of the most prominent challenges and strategies to accommodate the content. As with any area of mathematics, the TSVI and general education math teacher will need to collaborate often to ensure that the students with visual impairments have access to the curriculum. For the TSVI, their role and responsibility should focus on providing access while the general education teacher focuses on teaching the content. Working together and using appropriate strategies, tools, and accommodations, the students will have the same opportunities as their peers.

REFERENCES

Chang, L. A. (1983). *Handbook for spoken mathematics (Larry's speakeasy)*. Lawrence Livermore Laboratory.

Kaput, J. J., Carraher, D. W., & Blanton, M. L. (2008). *Algebra in the early grades*. Lawrence Erlbaum Associates.

National Council of Teachers of Mathematics (NCTM). (2000). *Principles and standards for school mathematics*.

National Council of Teachers of Mathematics (NCTM). (2018). *Catalyzing change in high school mathematics: Initiating critical conversations*.

National Governors Association Center for Best Practices & Council of Chief State School Officers (2010). *Common core state standards: Mathematics*.

Nemeth, A. (2013). *Mathspeak: A talk on verbalizing math by Dr. Abraham Nemeth, creator of the Nemeth math braille code*. https://accessinghigherground.org/handouts2013/HTCTU%20Alt%20Format%20Manuals/Math%20Accommodations/07%20MATHSPEAK.pdf

Van de Walle, J. A., Karp, K. S., & Bay-Williams, J. M. (2019). *Elementary and middle school mathematics: Teaching developmentally* (10th ed.). Pearson.

Warren, E., Trigueros, M., and Ursini, S. (2016). Research on the learning and teaching of algebra. In Á. Gutiérrez, G. C. Leder, & P. Boero (Eds.), *The second handbook of research on the psychology of mathematics education* (pp. 73–108). Sense Publishers.

CHAPTER 7

STATISTICS IN K-12 EDUCATION

Sean Tikkun

As a subject, statistics is the measurement and review of past events or data. The methods of statistics tell us whether an event is expected or unexpected through comparing the data to "normal" events. This comparative power of statistics makes it the backbone of almost all experimental domains of science and a key tool for understanding how complex systems behave, so it is vital preparation for students who are blind or visually impaired that want to enter STEM professions, the business world, or many other fields. Statistics is also used to present information about the world to the public by governments and news media, so a student who has a strong foundation of mathematical instruction in statistics is poised to be more informed and make stronger decisions.

Statistics is usually considered an advanced mathematics topic, but students encounter it throughout the early and middle grades long before they arrive in a dedicated statistics classroom. Students should already be comfortable with plotting points, graphing lines, and working with the slopes of lines from Algebra 1; working with curved functions and manipulating rational equations should be well developed in Algebra 2. During these courses, students should have learned to compute the basic measures of central tendency (mean, mode, median, and range), which are statistical skills. Students will have described data and referenced information from graphs and charts (such as bar graphs, pie charts, etc.) (NCTM, 2000). If these concepts are avoided or omitted because the instructors believe they are time consuming or cumbersome, this may set up the students to have a significantly larger challenge in later coursework (Kapperman, Heinze, & Sticken, 1997; Smith & Smothers, 2012). Statistics is not the most mathematically complex subject students face in their primary education, but it can still represent a significant jump in abstraction and complexity.

The study of statistics in K–12 education is intimately tied to the practice of reviewing data, particularly through data visualization; consequently, students with visual impairments will need specific accommodations throughout their instruction (Erhardt & Shuman, 2015). In statistics, data reviewing is greatly aided by tables and visual organization of the numbers that allow readers to quickly understand key information about the data (Stone, Kay, & Reynolds, 2019). When the data become more complex or large, specialized graphs are utilized to represent the data succinctly. The business and technical world calls these graphs "data visualizations," but in the statistics classroom we generally call them by their specific names. There are many graphs that help users understand different aspects of a dataset. The most common, scatterplots (Figure 7.1) and histograms (Figure 7.2), simplify and group data so it is easier to see patterns.

Figure 7.1 Tactually Produced Scatterplot

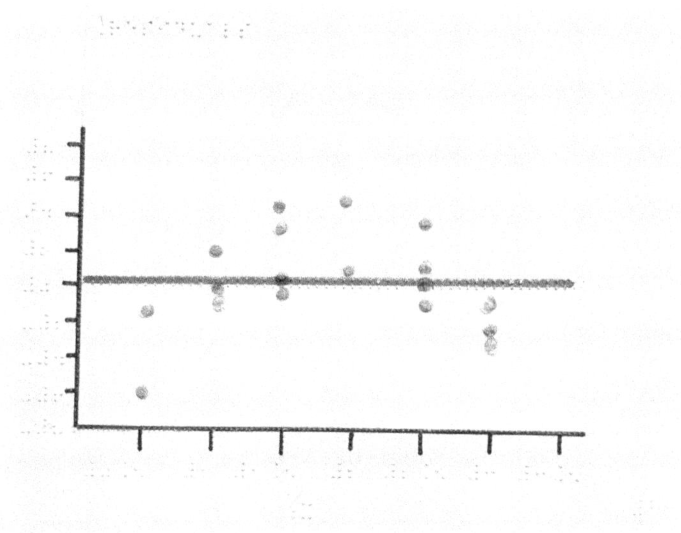

This scatterplot was produced using a tactile wheel and an embossing press and hammer and labelled with Nemeth braille.

Figure 7.2 Tactually Produced Histogram

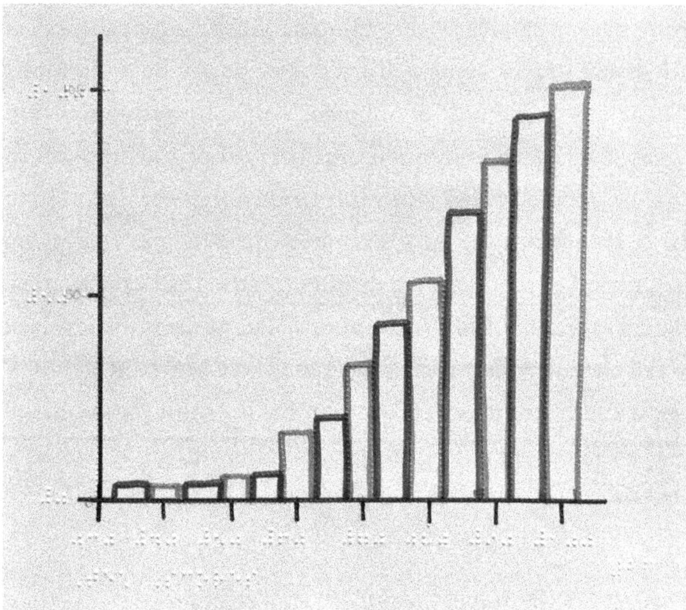

This histogram was produced using a fine tactile wheel with bars outlined in alternating red and green felt marker.

As the figures demonstrate, both of these methods of organizing data can easily be translated to tactile formats, but understanding these visual tools requires increased effort or cognitive demand for the individual with a visual impairment, as students who are blind or visually impaired often rely on part-whole understanding—meaning they must review parts of the graphic and then put the "pieces" together in their brains—when they apprehend a tactile graph (Kapperman, Heinze, & Sticken, 1997). Thus, it is often more difficult for them to use data visualizations to get a quick summary of a dataset, particularly as some data visualizations can be quite complex (Erhardt & Shuman, 2015; Abdalla, Alhaj, & Ramadan, 2025).

The tasks students encounter involving statistics become increasingly complex over time; however, like most mathematics, this complexity is overcome through repetition and practice. By the time students arrive in a statistics class in their final years of high school or early college, they should be prepared from their tactile literacy in previous courses (Rosenblum & Herzberg, 2019).

CONSIDERATIONS WHEN WORKING WITH DATA VISUALIZATIONS

The study of statistics focuses on working with datasets, typically in a table format or as a data visualization (i.e., graph). For small datasets, working directly with the dataset in a table format may be the easier route for basic statistical analysis. Teachers must consider a few concepts when faced with the prospect of teaching students with visual impairments to work with raw datasets in a table format. First, the students must understand the structure of tables and be able to read the data via print, braille, and/or a digital spreadsheet platform. Second, students with visual impairments must have appropriate experiences reading data presented in tables. This will include being able to use appropriate assistive technologies to access data via digital spreadsheets.

When datasets become too large or complex, advanced statistical analyses can be conducted using computer software applications (such as Microsoft Excel). This analysis can be aided by data visualizations, helping individuals see trends or patterns in the data. For younger children, data visualizations will help build familiarity and intuition about represented data. The fields of finance, epidemiology, and political science make increasing use of data visualizations to communicate information to the public. There are numerous ways to transform raw datasets into data visualizations directly from spreadsheet software packages that are outside of the purview of this chapter. Regardless of the specific software, though, the data visualization (i.e., "graphs"), the graphics that are created must be made accessible.

Preparing a dataset for presentation in braille should be done with consideration for the mechanics of reading braille. While printed data may be listed in multiple columns, searching multiple columns can be an added burden for braille readers. Providing students with both a working copy and a reference copy of the data will allow them to "mark" the data values as they transfer them to an ordered list. (This "marking" could be as simple as mashing the dots to note the value has been used.) If the students are using a tool such as a stem-and-leaf plot (see Sidebar 7.1), they can move a paper clip down the data list as they record numbers. This paper clip method will be useful for students with low vision, as well, to aid in the change of magnification or focus between materials. If students have access to spreadsheet software or a graphing calculator and data sorting is not the skill being developed or tested by the activity, they may use the spreadsheet's sort function to arrange the data automatically.

Access to these tools may be viewed as a reasonable accommodation, given the cumbersome nature of ordering data.

For students with low vision, the major consideration of any graphic will be related to the complexity and size of the parts. For example, a bar graph with three bars is simple as compared with a bar graph with 15 bars. When constructing the graphs, it is important that the lines are dark and clear, the background free of clutter (avoid using too many gridlines), and the colors distinct and have strong contrast. Graphs may use contrasting colors to make it easier to discriminate specific values or weights. Labels should be appropriately sized but permit enough space to be distinct from one another. More advanced data visualizations in upper grades may consist of digital formats on calculators or laptops, and so students with low vision will need to use appropriate assistive technologies skills to zoom in and out or leverage auditory information while staying oriented to the information.

For students with limited or no functional vision, the data visualizations include similar yet distinct challenges that need to be considered. When creating simple data graphics (such as a bar graph, pie chart, or histogram), the tactile graphics should follow specifications as outlined by the Braille Authority of North America (BANA). In order to differentiate parts of graphs, such as points, lines, and areas, different types of textures and shapes will need to be used. While creating tactile graphics can make many of the basic graphs accessible, they can become inaccessible as the dataset grows and more advanced data visualizations are used. For example, a scatterplot with 10 points easily can be made accessible. The teacher of students with visual impairments (TSVI) and math teacher will need to consider the key elements of the data visualization and determine how to present it based on the students' learning media preference and the specific task. For example, for a scatterplot with 1,000 points, maybe the tactile graphics will summarize the "shape" that the points make along with a trend line.

Unfortunately, more advanced data visualizations that are typically presented using computer software may not be accessible at the time of this book's publication. While there are ongoing efforts to make data science accessible to individuals with visual impairments (such as the Quorum Project, found at quorumlanguage.com, and SAS Graphics Accelerator, found at support.sas.com/software/products/graphics-accelerator/index.html), the emerging area of dynamic data visualizations promises new challenge in access. As our society becomes more data literate, the variety and combination of tools used will likely also become more complex, making the area of data accessibility a continuous area for development.

COMMON MEASUREMENTS IN STATISTICS

Measurements of Central Tendency

A core concept in the study of statistics—and the first method of understanding data in statistics—is finding the middle of a set of data (*central tendency*), but there is more than one center to a dataset. Mathematically, there are three ways of viewing the center of data: *median*, *mode*, and *mean*. Most people are familiar with these terms, but they may not have a clear grasp on what each one defines and how it is useful, as each measure gives a different approximation of center and describes the data in slightly different ways.

Median and mode will be prominent in first and second grade, though this terminology is not typically used until sixth grade. The counting of these two measures makes them good practice and application of skills learned early in math. Finding the mean, however, requires adding multiple numbers, possibly with multiple digits, then dividing that total by the number of items. Students lack the necessary skills until the third or fourth grade. After fourth grade, units on statistics will frequently use the mean more commonly but will still review mode and median.

Figure 7.3 Basic Histogram Showing Mean, Median, and Mode

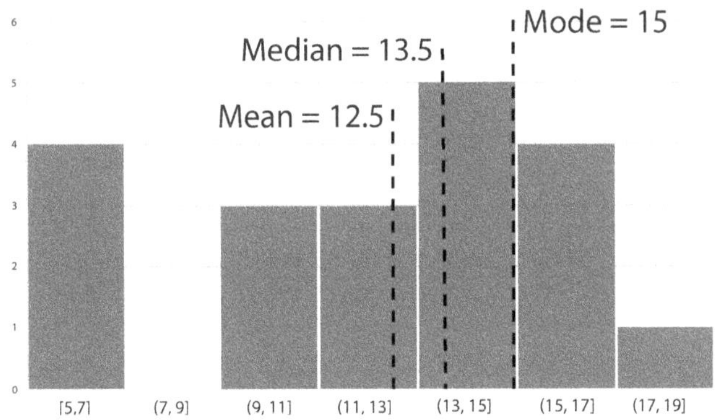

A histogram of a small fabricated dataset demonstrating the mean, median, and mode.

Median

The median is the "middle" of the data when it is arranged from the smallest to largest value. If a dataset has an odd number of values, it will be the middle value; if there are an even number of values, it will be the average of the two middle numbers. A median will present the center in terms of values in the dataset. As a descriptive statistic, it is useful when there are many terms at the high or low end of the dataset. For example, the median price of the cost of a wedding will have extremes that will impact a mean (average), so the median would be a better indicator of the "middle" value. The median will make the impact of outliers on the data apparent when compared with the other measures of central tendency.

Organizing data to find the median (and, if they are looking for more granularity, quartiles) will require the students to list the data points from smallest to largest. These datasets will typically reflect the order in which the values were collected, so they will seldom be sorted by value.

Sidebar 7.1 Using Stem-and-Leaf Plots for Data

A stem-and-leaf plot is a special table that typically has a column with the first number (stem) and a column with a "leaf" that represents each number in the dataset. The stem will normally be the tens or higher place value, and the leaves will represent the next place value to the right. For example, if we were organizing the number 25, the 2 would be the stem and the 5 would be the leaf. For a larger value such as 257, the 2 might be the stem and 57 would be the leaf; alternatively, the 25 might be the stem and 7 would be the leaf. These choices are dependent on the range of data, but the "leaves" should be put in increasing order, allowing students to succinctly show the data in lowest to greatest order.

The dataset 25, 34, 16, 15, 18, 25, 39, 22, 25, 9 can be represented in a stem-and-leaf plot:

Stem	Leaf
0	9
1	5, 6, 8
2	2, 5, 5, 5
3	4, 9

Sidebar 7.1 *(continued)*

Students can take the raw data from the "reference page" and create a stem-and-leaf plot on their "working page." If the data are listed on the reference page, students could either "erase" the data points as they add them to their table or they could use a paper clip to maintain their place in the list of the data.

Stem-and-leaf plots also have the utility of being organized vertically. This makes it much easier to organize using a braillewriter. A well-organized stem-and-leaf plot with ascending values might require two passes: the first pass to organize the data by stem, and the second to put the leaves in order.

The result of the stem-and-leaf plot also has the shape and characteristics of a histogram. The median and mode will be much easier to find in the organized lines of data than in a long horizontal or vertical list. This ease of use is also true for finding interquartile ranges.

Mode

Mode is the most frequently occurring data value, which is a valuable statistic in both population and probability work. It is very valuable in studying populations, as it can identify pluralities or majorities in a dataset (the most common blood type, for example). It is also used in probability work, as it can identify the most likely outcome of a given scenario. When viewing data visualizations, the mode will be readily visible as the peak in a histogram or a dense collection of points in a scatterplot. As the most frequently occurring value, it provides immediate reference for the most common result of a test or the most common trait in a population.

Finding the mode of a dataset is made much easier if the dataset is ordered. Each value will require a count with the highest count being the mode. Reviewing a dataset in order should provide students with a rough estimate of the largest counts. Making use of a stem-and-leaf plot can be useful for showing both median and mode. It can also serve as a highly accessible histogram that preserves the dataset values.

Mean

The mean, or average as most people think of it, is the middle defined by adding all the points and dividing by the number of points. The mean provides the center of the data as mathematically derived. As the mean is easily represented mathematically, it is the easiest of the central tendencies to manipulate for further analysis. It is also the measure that is most susceptible to data featuring unusual peaks or significant outliers (extremes). A significant amount of effort in statistics is required to ensure that the mean is a trustworthy measure.

Finding the mean is far less cumbersome than the other two measures of central tendency. The values can be added in any order and divided by the number of terms. Nevertheless, the paper clip method of keeping track of entries is still useful, and entering the data from an ordered set can make the process faster.

Combining Central Tendencies

Taken individually, each of these values gives us a specific descriptor; observed together, they can describe the overall feel and character of a set of data with just three numbers. It is the hope of math educators that students will generate an intuition based on experience with these measures of central tendencies.

For students with visual impairments, understanding the importance of these numbers and obtaining proficiency with finding them should not pose a challenge, but instructors should be prepared to provide materials for tactile graphics. Without the measures of central tendency accompanied by histograms or box-and-whisker plots, students will not develop the same mathematical intuition as their peers. Spreadsheet software or the TI84-Orion Graphing Calculator (orbitresearch.com/product/orionti-84-plus) will provide the three measures of central tendency with a simple command, but it is not unusual for teachers to limit these tools while students learn the routines of finding the values.

Quartiles and Percentiles

The second method of understanding data in statistics is by grouping the data into quartiles. In the same manner that the median finds the middle term in the data, quartiles divide the data into two additional halves on either side of the median. Most people will be familiar with the term *percentile*, often used with standardized scores and growth measurements in children.

For very large datasets, such as populations in the hundreds of thousands, percentiles break the quartiles down even further. While quartiles are normally used to discuss wherein the dataset the point exists, percentiles describe how much of the data (population, for example) is below the specific data point. Both of these descriptors give a quick "feel" for the location of the data. The actual "cut points" on the quartiles also communicate the skew or distribution of the data. Skew will be discussed later in the Sampling Distribution section.

GETTING STARTED WITH REPRESENTING ONE- AND TWO-VARIABLE DATA

The first activities with data are introduced in the early elementary grades alongside skills such as counting, and they typically begin with a single, simple variable. This typically begins with having students count objects and then maintaining "counts," such as with tally marks. Early-grade graphs and charts frequently display data about candy color, height, eye or hair color, and preferences (favorites). One of the earliest forms of a data graph is a *pictograph*, where students place actual objects on the graph to represent counts. Initially, the objects (or stickers and, later, graphics) represent data on a one-to-one ratio; over time, the objects can be set to mean more than one object (such as one sticker represents three people). A key (or legend) may show that each object counts as two, five, or 10 of the objects; in these cases, partial objects may be used to show when there is less than the total, like half a soccer ball representing only one person rather than the two a whole soccer ball would represent.

The next graphs students commonly encounter are *bar graphs* with single categories. The pictographs created by students can be converted to "bars" using manipulatives such as Unifix Cubes (didax.com/unifix-cubes.html), then activities can allow the students to compare the "bars" formed by the cubes. From the manipulatives, the graphs can then be transferred to a picture of the graphic. As part of the lesson, students will need to become familiar with the most common parts of a graph. For example, students need to learn to read the title of the graphic, find and read the labels for each axis, look at the intervals denoted by tick marks, and then read the bar graph.

In the upper elementary grades, students receive experiences with more detailed bar graphs, including scales, multiple bars, and advanced labeling that provide more information. *Pie charts* are often presented around the same time as bar graphs, as they can be used to interpret proportions and estimate percentages by reviewing the categories of the data. Around the same time, students will be introduced to *line plots* (i.e., *dot plots*) that show simple numerical data as stacked "x's" or "dots" on a number line, similar to pictographs.

In middle school, the statistical graphs become more complex as students learn to use different types of graphs to interpret different types of data. Students are introduced to *histograms* (which they will recognize as similar to bar graphs but with numerical data grouped into intervals called "bins"). More advanced graphs (discussed later), including *box-and-whisker plots* (or *box plots*), *stem-and-leaf plots*, and *frequency distributions*, are introduced.

In high school, the types and complexity of statistical graphs can grow exponentially, depending upon the mathematics curriculum. Students will perform regressions with scatterplots and review time series graphs (which show data trends over time); they also will be introduced to box plots (with outliers), frequency polygons (which are similar to histograms but show frequency trends), and advanced data visualizations.

Adapting Statistics Graphs

There are many considerations when adapting statistics graphs for students who are blind or visually impaired, and TSVIs should keep the ultimate goal of representing the data in mind. Consider the pictographs in the previous section: In converting these pictographs for students with blindness or visual impairments, TSVIs may be tempted to use similar illustrative objects, such as a soccer ball or specific symbol for the pictograph. In these cases, it is important to remember that the *chart* is the focus of the learning activity. If the TSVI or the students spend more time on the illustrative object than on what it represents, then it diminishes the purpose of the chart. When making tactile versions of these types of visualizations, simple substitutions are recommended, such as craft foam circles or squares. The pictographs can be created entirely in braille as well, using the lowercase "L" character for half and "for" for full. In these cases, it is recommended to lay the pictograph out horizontally rather than vertically so it is more comfortable for the student to read and label (BANA, 2022). Line graphs can also be used; the skills and methods for creating them are no different from what would be used in algebra or other Cartesian graphing (see Chapter 6, "Foundations of Algebra").

For more advanced data visualizations, the Guidelines and Standards for Tactile Graphics (BANA, 2022) sets out clear and effective means to adapt the graphics that most frequently appear in K–12 statistics. The largest concern is that the methods used in the modern classroom will frequently require students to produce these graphs on demand. (For example: Students are asked to count or measure things in class, such as the number of candies of a particular color or the height of each student. This information, or novel data, collected in class will then be made into a graph. At early grades, this will likely be a pictograph—potentially made out of the candy itself.) Some websites, such as Khan Academy and IXL Math (khanacademy.org and ixl.com/math, respectively) will provide clickable or draggable icons. At the high school level students, may enter the data into a spreadsheet on either a graphing calculator or a computer. The tool will then instantly produce a graph.

Mathematics educators may take this ease for granted without realizing that no similar, speedy process exists to make these graphs in a tactile format as quickly and easily. As ease and speed are barriers, it is necessary to prepare classwork ahead of time. In the case of novel datasets, tools such as an All-In-One Board from the American Printing House for the Blind, or APH (aph.org/product/all-in-one-board) with prepared objects or bars will allow for quick design. Bar charts and histograms can be prepared with square magnets that can be stacked to create a solid bar. Stencils and templates can be developed out of cardstock for repeated tasks, such as drawing axes with evenly spaced markings.

Figure 7.4 Stencils Made of Braille Book Cover Plastic

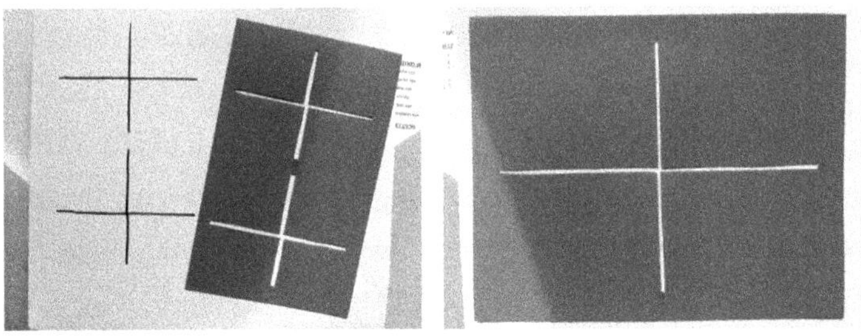

Grid templates can be created using flexible plastic braille book covers and cut out with scissors or a utility knife.

These graphs will also require orientation, tactile literacy skills, and practice for students to become proficient in finding important features during class. In later coursework, such as in Advanced Placement (AP) and college statistics classes, students will learn to use programs such as SPSS, SAS, R, or other statistical software packages to produce their graphics. SAS's Graphics

Accelerator browser add-on allows for the sonification of graphs (see Sidebar 7.2). The Quorum programming language also offers students the ability to create graphics through basic programming.

Sidebar 7.2 Sonification

Sonification has been a novel solution to graphs for roughly 20 years. In that time, very little has changed about the theory or approach, but the tools have become more widely available, including web tools to easily integrate them.

Tones are used to communicate height on the graph. Higher tones communicate larger values, and lower tones communicate smaller values. These tones are present to communicate the location on the y or vertical axis. The traditional method of communicating the x or horizontal axis is over time. If a student plays a graph, it starts from the left side of the graph and makes tones communicating the values from left to right. As a result, time becomes a kind of adaptation of the x-values. If an individual navigates the data freely, using the left and right arrow, they can differentiate whether values are getting larger (positive slope), smaller (negative slope), or staying the same (0 slope). Students and TSVIs can review this using the online graphing calculator by Desmos or an Orion TI-84.

- With Desmos: If a graph is present, select the audio tracking keyboard by pressing Alt-T. This feature was designed to supplement the use of a screen reader but is testable without. (I recommend typing $y = sin(\theta)$ or $\sum_k \binom{n}{k}$ to get a nice feel for the sounds.)
- With Orion: When a graph is present, the calculator will provide sonification when the "Trace" key is pressed. Moving left and right will give a relative tone, depending on the position on the grid.

Sonification can give a user the feel for the shape and changes in a graph in the same way glancing at the graph for a person with sight will. To truly review the graph, the user will need to have a firm understanding of the axis measurements, the coordinates of the high and low points, and the coordinates of any intercepts with the axis.

The ability to produce accurate and usable alternatives for the graphs in statistics does not mean that the teaching of these graphs will always directly translate. Students will still need to be introduced to each of the types of graphs described below in advance of coursework. Components of each graph and their critical features will need to be defined and explored with examples that allow the students to draw meaning. The students will then need to learn to find the labels, trace the axis values to the graph features, and maintain orientation while exploring with the right and left hand. Graphs will occasionally have a key that references the equation or dataset for a specific line or dot type; reading these keys before delving into the graph analysis will also be important. Because statistics relies upon analysis and a certain intuition about data, the students will need to access the datasets at the same time as the graphic to build this understanding.

Bar Charts

Bar charts are used across curricula and retain their usefulness even in college-level statistics courses. These charts can be used to represent a variable for single or multiple groups. They can also chart multiple variables. As mentioned earlier, these charts will be introduced in the early elementary grades and gradually grow more complex.

Bar charts are normally used to plot data that falls into categories. A category could be a color, a year, or a specific score or measurement. These categories are then counted, and the bar height reflects the number of items in the category. This can also be thought of as events or how often an event happens. An event example would be the number of days with a temperature over 90 degrees, or the number of home runs hit by a fictional baseball team. In the early grades, symbols or pictures may be used to create a pictograph as a precursor to bar charts. These symbols will sometimes reflect more than one event, with partial symbols showing a fraction of the amount in the key. In the example below, a baseball equals 2 home runs and the half baseball equals 1 home run. (The unique symbol is less important than the fraction concepts that are being reinforced, although these symbols can be opportunities to explain context for more general awareness.) APH MathBuilders: Statistics kit (aph.org/product/mathbuilders-statistics) includes plastic molds and accompanying materials that can be used to create pictographs and/or bar graphs.

Figure 7.5 A Simple Pictogram Bar Chart With Symbols

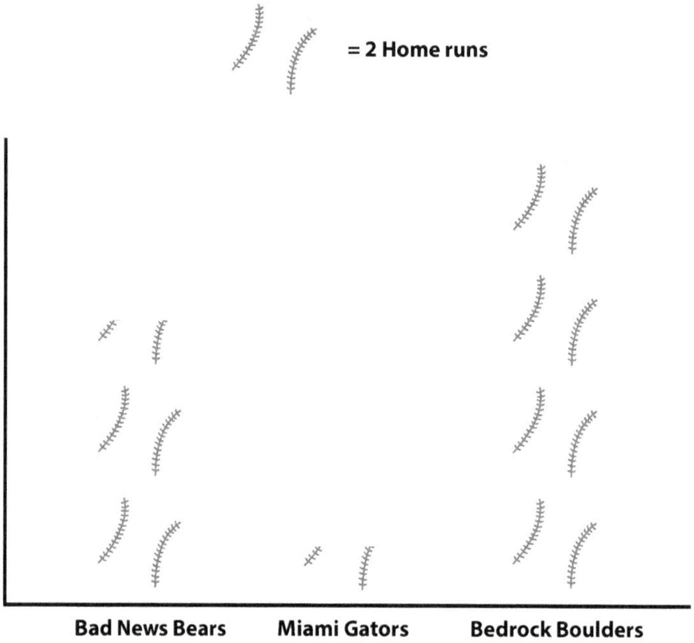

Simple pictographs are used in the early grades and may use objects instead of simple shapes to represent counts. In this example, one baseball is equal to two home runs, and so the image shows half baseballs that will represent one home run.

When introducing a bar chart, the title and axis labels should carry the critical information for the data: The title tells the reader what is being described, and the axis lets them know what the variables are and how they are measured. Bar charts can be represented either vertically or horizontally. In a vertical bar chart (as in the pictogram bar chart in Figure 7.5), the x-axis will be the categories, and the y-axis will be the counts for each category; in a horizontal bar chart (see Figure 7.6), the x-axis will be the counts and the y-axis will be the categories. In order to read the counts, any tactile or modified chart will need guidelines to ensure the reader can connect the measures on the y-axis on the left to the bars, even as they get farther from the left side of the chart.

Figure 7.6 Horizontal Bar Chart Showing the Vision-Impaired (VI) Child Count for Southeastern States

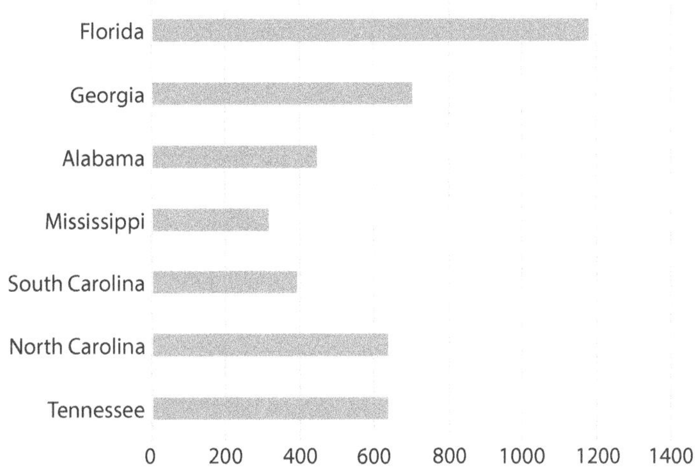

This horizontal bar chart uses the horizontal axis to represent the number of children who are blind or visually impaired in each state.

Adapted from EdFacts Data Warehouse (EDW), 2020

Typically, teachers will ask students to identify the biggest or smallest group. The class may be asked to identify the count of a specific group, or all groups with a count above or below a certain value. When completing these activities, the students will need to reference the axis each time. This need for reference and clarity is no less important for students using tactile materials than it is for those with low vision. The Draftsman Tactile Drawing Board from APH (aph.org/product/draftsman-tactile-drawing-board) and TactiPad drawing board from Thinkable (thinkable.nl/tactipad) are both useful tools for these activities. Both devices have an adjustable ruler that can be moved and locked to mark a location or provide reference along an axis. The Graph-Grid application for the TactiPad can also offer multiple adjustable references to aid in tracking lines and maintaining orientation.

An alternative to these tools includes a good-sized ruler (or three) with putty adhesive on the back that can easily be affixed and removed. The added benefit of the putty is that it lifts the ruler to decrease the risk of crushing or tearing the embossed information. When using enlarged materials or magnification, students will have a restricted field that will likely not include the axis. Using a ruler on physical materials offers some benefit over digital magnification that might make maintaining a reference challenging.

Histograms

Histograms appear similar to bar charts and have similar features when it comes to reading and identifying information, but they ultimately present different information. Histograms differ visually in that they do not have spaces between the data bars. This lack of space reflects the fact that histogram bars measure data that falls into a range. Whereas a bar chart would measure a category (e.g., the number of people who like chocolate ice cream), the histogram bars tell the reader how many points fall into a range of measures (e.g., the number of points between 100 and 200). Histograms also do not normally describe categories but rather data along a range of values.

For example, the bar chart of the number of children with visual impairments by state in the Southeast (United States) region (Figure 7.6) has very specific defined categories. In the bar chart, each bar will represent a state and a count. The histogram of children with visual impairments counts by state (Figure 7.7) has columns that represent a count of 1–100, going up by 100 for each column. If a state has a count in that range, it is placed in that group.

Histograms are read from left to right and will typically have a peak at some point. Data will often taper off to the right and/or left of the peak (which does not center on the mean value); however, sometimes the data does not follow typical patterns. In Figure 7.7, which depicts children with visual impairments counts by state, the peak is in the 0–100 range; this reflects the large number of smaller states, as California and Texas were omitted as outliers. Gaining and understanding the range of the groups is important before reviewing the shape and distribution: Two histograms of the same data with different ranges can look very different, as is demonstrated in the second histogram, which employs ranges of 500 instead of 100 (Figure 7.8).

Figure 7.7 Histogram Showing the Vision-Impaired (VI) Child Count by Number of States (100 Range)

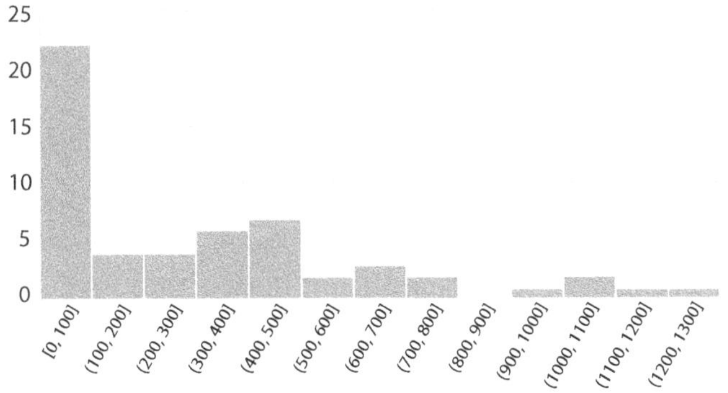

In this histogram, each bar vertically represents the number of states with a student count between the two values listed on the horizontal axis.

Adapted from EdFacts Data Warehouse (EDW), 2020

Figure 7.8 Histogram Showing the Vision-Impaired (VI) Child Count by Number of States (500 Range)

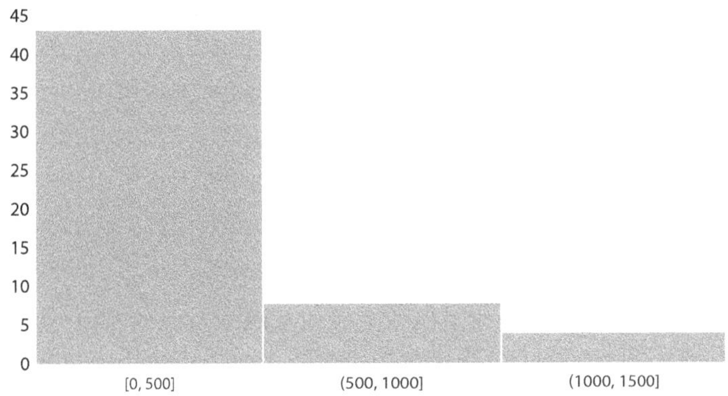

Changing the range of a histogram can provide a different level of nuance in how the data is presented.

Adapted from EdFacts Data Warehouse (EDW), 2020

Box Plots

A box plot, sometimes called a box-and-whisker plot, is simply five data points presented in a horizontal or, less commonly, vertical line. The five points are the median, the first quartile, the third quartile, the minimum, and the maximum, and each is plotted with a line at the values.

- The *median* is normally bold or a heavier weight line.
- The *first and third quartile* are standard lines that are the same length as the mean line.
- The lines representing the *extrema* (*minimum* and the *maximum*) are the same weight but tend to be half or a third of the length.
- Finally, lines are drawn to form a box between the two quartiles, with the *median* inside the box.

This plot presents the critical information that describes the dataset, laid out visually. While this same data could be presented in a table, the spatial distance between the median, quartiles, and extremes would be lost; further, using a table also erases the ability to compare multiple datasets over the same range and measurements in datasets (see Figure 7.9).

Students should be introduced to box plots at the same time as the terms *quartile* and *extremes* are introduced. Language around relative distance and spatial orientation should be used to describe what the students see, whether it is tactile or visual. Statistics relies heavily on descriptive language, which is one reason why the graphs and values are called *descriptive statistics*.

Box plots carry a wealth of information compared with histograms or bar charts. The primary benefit of a box plot is the fact that it presents the quartiles and extremes of the data without the need to scan through a large data table or follow the height changes in the bars. A box plot provides the minimum, first quartile, median, third quartile, and maximum values as a summative graphic of the data. The function of the graphics is to provide information related to skewness and symmetry, identify outliers, compare across groups with side-by-side comparisons of multiple box plots, and quickly determine spread and central tendency.

Students with visual impairments will need access to high-quality graphics that provide access to the data contained within the dataset. This would involve including all parts of the graphic and ensuring that applicable labels are provided.

Figure 7.9 Sample Box Plot

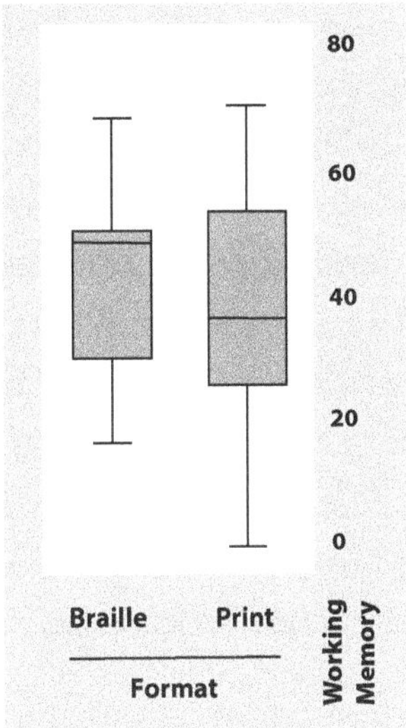

These box plots comparing the working memory of college students who use braille or print quickly show the differences in the data between the two groups.
Adapted from Tikkun, 2018

Figure 7.10 Tactually Produced Box Plot

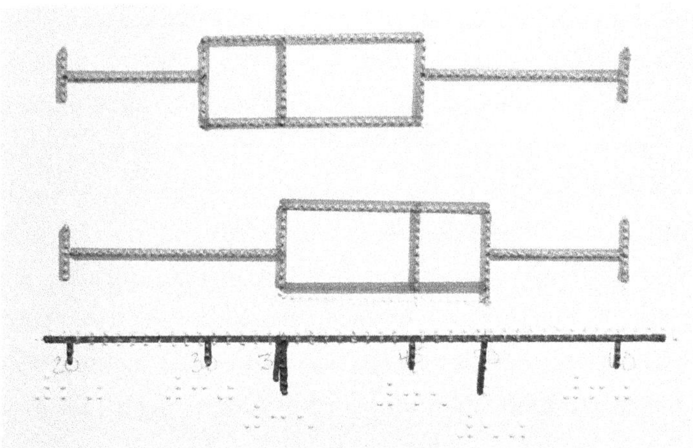

Two side-by-side box plots in a tactile format, laid out horizontally. All lines of the boxes were made using a fine tactile wheel.

Scatterplots

Creating a scatterplot is similar to the standard graphing of an algebraic equation. Instead of starting with an equation to define the points, scatterplots map the data as points, so it is laid out spatially upon the Cartesian graph (x–y graph). The purpose of a scatterplot is to review data holistically across two variables to determine any possible trends. Students will use this data to find an equation of a line, called the *line of best fit*. (Please note that advanced statistics may present situations where there is a curve of best fit instead of a straight line or no determinable line of best fit, but students are unlikely to encounter this in K–12 coursework). It is easiest to describe the process as the reverse of what students do in algebra classes: In algebra, students use the function to determine the points along a line; in statistics, students use data points to determine the function of the line.

For students who are blind or visually impaired, the challenges of working with scatterplots are largely associated with the crowding of data. Closely graphed data can make it impossible to distinguish the density that questions frequently reference. For example, a scatterplot may have 1,000 individual data points (dots), which would mean very little in the form of a tactile graphic. Therefore, instruction on scatterplots should start with only a few data points before moving forward with larger datasets. As with many statistics graphs, a sorted data table will help clarify details when data are densely packed. When students work on finding the line of best fit, they will use data tables, calculators, and statistical software programs.

The guessing, or estimating, of the line of best fit is a less important skill now that computers can provide it. The most important feature in these lines is always the slope. As a result, it is critical students know how to find the slope of an equation, and all adaptations should include the equation of any lines present. This relationship will lead to a statistical measure called the *correlation coefficient*, which bears similarity to the slope but is a rather complex equation related to covariance and standard deviation. The understanding of lines and slopes is the critical foundation for students interpreting graphs in statistics. The process and methods of creating accessible versions are no different than the methods used in creating accessible standard graphs in algebra.

Figure 7.11 Sample Scatterplot

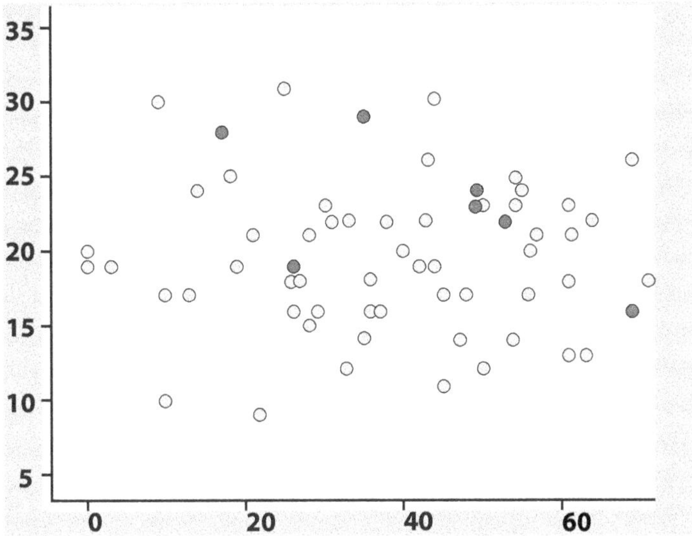

The plot shows two sets of data: a large number of open light blue points denoting a control group with no clear grouping or relationship, and a smaller number of closed red points denoting the comparison group with a slight negative relationship.
Adapted from Tikkun, 2018

PROBABILITY

As a subject, probability is tied to statistics in that the same theories inform each domain, but probability studies future possibilities and statistics studies current or past events to understand occurrences or to help make predictions. In K–12 education, probability tends to be engaged in a small number of variables with known and defined outcomes (Sidebar 7.3 demonstrates basic probability using 6-sided dice). It can, however, chain events and variables together to give likelihood for complex situations. For example, weather predictions rely on complex probability models. In any case, determining probability requires one to define the variables and plot the likelihood of different events. Similar to statistics, the process of organizing results and describing event chains is aided by spatial organization and graphs.

Sidebar 7.3 The Probability of Different Values When Rolling 6-Sided Dice

Most students will be familiar with rolling dice to play games. The below probability table shows the likelihood of the events of one 6-sided die (1d6), two 6-sided dice (2d6), and three 6-sided dice (3d6). As they add dice, the resulting probabilities begin to represent a smooth progression up and down. This situation can be further explored by playing a game such as Monopoly with one, two, or three 6-sided dice. Students can see how many roles it takes to pass "Go" and develop expectations of how the game might progress. The 2d6 scenario is the basis for the game of craps, where a number is rolled and a player needs to reroll it before rolling a 7 (the most likely number). Without betting, it can be fun to see how many rolls or wins one can achieve before the 7 appears. More recent games, such as Settlers of Catan, require players to understand probability in order to keep pace in the game. Accessibility kits for many games can be purchased from 64 Oz. Games (64ouncegames.com).

Result	1d6 probability	2d6 probability	3d6 probability
1	1/6 = 16.67%		
2	1/6 = 16.67%	1/36 = 2.78%	
3	1/6 = 16.67%	2/36 = 5.56%	1/216 = .46%
4	1/6 = 16.67%	3/36 = 8.33%	3/216 = 1.39%
5	1/6 = 16.67%	4/36 = 11.11%	6/36 = 16.67%
6	1/6 = 16.67%	5/36 = 13.89%	10/216 = 4.63%
7		6/36 = 16.67%	15/216 = 6.94%
8		5/36 = 13.89%	21/216 = 9.72%
9		4/36 = 11.11%	25/216 = 11.57%
10		3/36 = 8.33%	27/216 = 12.50%
11		2/36 = 5.56%	27/216 = 12.50%
12		1/36 = 2.78%	25/216 = 11.57%
13			21/216 = 9.72%
14			15/216 = 6.94%
15			10/216 = 4.63%
16			6/36 = 16.67%
17			3/216 = 1.39%
18			1/216 = .46%

Random Events

The random variables of probability center around events, which are recorded as fractions or ratios. Statistics treats several mathematical symbols and methods differently than students have encountered to this point. The familiar letter *m* from algebra generally represents slope, but statistics students will encounter a range of *m*s in size and style—sometimes with subscripts—representing different measures of central tendency. The use of a bar over a letter represents a line segment in geometry, but it will be used over letters to represent the mean of a data group in statistics. In statistics, additional characters might appear above letters, and Greek letters will be used frequently. This will make the initial introduction to the topic challenging, as the students are reading and hearing different terminology than they are used to.

This terminology and the writing of probabilities will not show up frequently in the K–8 mathematics curriculum; in the early grades, counting events and writing ratios is the extent of the work. The intention is to build a student's intuition around random events and establish a difference between expected and observed events. This is often done by flipping a coin, rolling dice, or drawing colored objects out of a bag. In each of these activities, there will be accessibility issues, but TSVIs and students can overcome them easily.

- Coins: Flipping a coin can be done by spinning it on its side or dropping it into a flat-bottomed container such as a box or a can. To ensure the coin's faces are necessarily distinct, use plastic coins (available at most dollar stores) and add tactile stickers as appropriate.
- Dice: Accessible dice are available with both embossed and recessed numbers or braille.
- Colored objects: These objects cannot be tactually discriminated, lest the results be skewed as students are choosing them. For this reason, colors should not be a distinguishing characteristic in objects for students who would need adaptation.

Another common probability activity is that of determining the odds of a game spinner pointer landing on a specific section of a circle subdivided into sections. While this tool can be used by students with low vision, completing tasks with braille and/or tactile stickers is more challenging. The APH Math-Builders Statistics Kit (aph.org/product/mathbuilders-statistics) includes a modified "spinner" where the circle spins and has a tactile arrow that selects where the spinner "lands" (such as the wheel on the *Wheel of Fortune* game show). The kit includes circles that are subdivided into fractional parts that allow exploration of different sections.

Notetaking skills are very important in recording random events. This is not only an easily accessible task in group work but also builds students' ability to scan and read data they have collected. While most students and teachers may initially consider having the student record on a braillewriter, the slate and stylus can be much more efficient tools that can be more easily read in the moment. (It is important to emphasize to students the difference between shorthand in notes and writing math for others to read.)

Another benefit of the slate is that there are markers on every fifth cell. In the case of tracking heads and tails, two slates can be used and easily managed. A Cranmer abacus also is a convenient tool for counting and tracking totals in this way, and the addition of rubber bands to section off the abacus could be helpful for multiple results.

The process of making tally marks in print typically uses a horizontal line on the fifth mark that crosses the previous four vertical marks. This is replicated in Unified English Braille (UEB) Math/Science braille using dots 456, the crossing mark is dots-25 or just having five dots 456 for a set in Nemeth code. Groups of tally marks are always followed by a space so readers can clearly distinguish groups. The process of making a tally mark and producing a three-dot character are not equivalent. Students may experiment with other methods, such as a single dot-1 for the ones and dots-24 or dots-246 for every fifth, which can be more quickly written and read.

Reading Event Probabilities

In a study of probability, the fraction $\frac{1}{4}$ would not be read as "one fourth" as it is in a typical math lesson. Instead, when working with probabilities, it would be read as "one in four" or "one out of four" events. (This can also be written as the ratio 1:4.) The ratio notation will be useful in K–8 study of probabilities, but the fractional form will be more common when discussing multiple events are used. This is also critical when dealing with compounding events, such as flipping a coin twice and seeking the same result (tails, then tails again).

Probability Distribution

Recording and charting the results of some random event will normally entail using a bar chart. As the specific event that occurs each time (heads or tails, in the case of flipping a coin) is a category, it can be presented with a bar for each event. For something like flipping a coin, the bars will trend toward the same height over time, demonstrating that each event is equally likely. When charting out a slightly more complex situation, such as rolling two 6-sided dice, a curve up and down (resembling a mound or hill) may emerge, but not always. The understanding that each event is independent and anything can happen is learned through sampling random events and displaying them graphically.

Sampling Distributions

Sampling distributions could easily be overlooked for their similarity in form and construction, but the interpretation and exploration are particularly valuable for learning and real-life applications.

Once data are collected, statistics is primarily concerned with examining and describing datasets. Most of the descriptions are tied to the *normal curve.* The normal curve is an ideal distribution, with a strong grouping in the center (the normal) and tapering data on each side of the central data. The histograms that students learn to use over time eventually result in a graph of a smooth curve that describes the distribution of many samples. Sampling distributions are a calculated statistic but are also used in probability predictions. The graph is a curve and the axes are labeled when values are less than 10. Nothing is particularly novel about the graph of a distribution, but a student's ability to estimate the mean and the standard deviation from the graph will be an important skill, as it forms a foundation of evaluating distributions and gaining quick understanding of the underlying statistics. Additionally, students taking the AP Statistics course may be asked to match descriptive statistics to graphs or vice versa.

The use of sonification with stepwise tracing will be very useful to students who continue in areas that depend on statistical analysis. The rising and dropping tone and the peak tone will enable students to get a feel for the shape, and then reviewing the points with tones will assist them in putting values to the tones. A "play" of the graph may be too swift to differentiate the height or slope, but tapping through points at a measured pace will allow for a more careful review. Think of the process as playing a song at a higher or lower tempo—it is easier to discern the individual notes more clearly at the lower tempo.

SUMMARY

The statistics and probability topics covered in this chapter are those foundations most commonly covered in the K–12 curriculum, introducing students to a rich domain of mathematics that is expanding with new predictive and dynamic models to this day. Higher level analysis and specific tests for conditions are not included, as they are simply equations and reported values. There is nothing unique about the method of reading or writing, given the students have learned the appropriate mathematical code or markup language.

The goal of K–12 statistics is to ensure that students have a foundational understanding of inferential statistics that starts early and continuously grows over many years. Ultimately, students can develop this understanding and put it to work in their lives and careers, potentially with far-reaching effects. During the Space Race, complex algebraic computation and calculus were critical to problems facing nations competing to land a manned spacecraft on the moon; today, statistics has more diverse applications as we strive to balance our finite resources with the demands and needs of people and planet. To do this, we require large amounts of data and carefully calculated expectations. The citizens of our near future must be literate in these methods so they can evaluate whether to trust or be skeptical of predictions.

Almost all scientific domains—not to mention finance and economics—rely on statistics. Even for students not pursuing a career in science or mathematics, they will still utilize the core skill of inference—the ability to make reasonable predication based on data and trends. A student using a cane to make a street crossing is accessing the same applied skillset, albeit in a very different—but no less important—situation.

REFERENCES

Abdalla, S., Alhaj, A., & Ramadan, E. (2025). Motivational factors for visually impaired college students in learning data analytics. *Eurasia Journal of*

Mathematics, Science and Technology Education, 21(2), em2578. https://doi.org/10.29333/ejmste/15923

Erhardt, R. J., & Shuman, M. P. (2015). Assistive technologies for second-year statistics students who are blind. *Journal of Statistics Education, 23*(2). https://doi.org/10.1080/10691898.2015.11889733

Kapperman, G., Heinze, T., & Sticken, J. (1997). *Strategies for developing mathematics skills in students who use braille.* Research and Development Institute, Inc.

Spinczyk, D., Maćkowski, M., Kempa, W., & Rojewska, K. (2019). Factors influencing the process of learning mathematics among visually impaired and blind people. *Computers in Biology and Medicine, 104*, 1–9. https://doi.org/10.1016/j.compbiomed.2018.10.025

Stone, B. W., Kay, D., & Reynolds, A. (2019). Teaching visually impaired college students in introductory statistics. *Journal of Statistics Education, 27*(3), 225–237. https://doi.org/10.1080/10691898.2019.1677199

Tikkun, S. R. (2018). *A study of the factors influencing algebraic problem solving by individuals with visual impairments* [Doctoral dissertation, Northern Illinois University]. Graduate Research Theses & Dissertations. 726. https://huskiecommons.lib.niu.edu/allgraduate-thesesdissertations/726

U.S. Department of Education, EDFacts Data Warehouse (EDW). (2020). *IDEA Part B child count and educational environments collection, 2019–20.* [Dataset.] https://data.ed.gov/dataset/71ca7d0c-a161-4abe-9e2b-4e68ffb1061a/resource/c49009eb-a269-4131-9bbe-7d8a3f67f649/download/bchildcountandedenvironments2019-20.csv.

Wild, T., Smith, D., Kelly, S., & Fast, D. (n.d.). The National Research Agenda for Science, Technology, Engineering and Mathematics (STEM) for Students with Visual Impairments. Material based upon works supported by the National Science Foundation HRD-2032668. https://u.osu.edu/nationalresearchstemvisualimpairment

CHAPTER 8

THE IMPORTANCE OF TACTILE GRAPHICS IN MATHEMATICS

Derrick W. Smith and Tiffany Wild

If a student opens any mathematics textbook, from kindergarten to calculus, they will find many commonalities. Inside are myriad numerals, text, and specialized math symbols. They will also almost always find graphics. Though most mathematics is possible without graphics, there are aspects of mathematics that require the use of graphics or accentuation by them. For example, much of geometry taught in schools (primarily Euclidean geometry) extensively uses graphics to support conceptual development. Another example, common in algebra, focuses on functions found within Cartesian geometry (or coordinate geometry; see Chapter 5, "Geometry"). While an algebraic function is essentially an equation, a student can use the graphical representation of the equation to graphically present all possible solutions of the function. In statistics, there is a strong emphasis on different types of graphics that provide a visual display of data for efficient consumption and interpretation; a well-rendered graphic can convey hundreds or thousands of data points clearly and efficiently.

Between the graphics found in mathematics textbooks that complement conceptual understanding and the applications of graphics in mathematics necessary to convey subject mastery, there is an urgent need for *all* students, including all students with visual impairments, to receive high-quality graphics that can strengthen their understanding of mathematics concepts. As such, teachers of students with visual impairments (TSVIs) should expect to regularly implement adaptations that ensure their students are able to access the graphical material in a mathematics lesson. To make graphics accessible for students with visual impairments, there are four main types of accommodation based upon the unique needs of the students and the purposes of the graphic connected to the content: enlargement, conversion to text, sonification, and tactile graphics.

Accommodation 1: Enlargement

For students with low vision, the most frequently used accommodation involves enlarging the graphics or providing magnification technology so the students may access the graphic. In this accommodation, the TSVI will need to ensure that the graphic is large enough to include all important information. They will also need to ensure no information is lost or changed during enlargement: For example, if the original graphic uses color to differentiate between items, then making an enlarged graphic in black and white would not be appropriate. If a student receives an image that they need to measure using a copy of a ruler that was enlarged from the original, then their answer will be inaccurate. In all cases, the TSVI should be aware of what information the original graphic presents and what it is asking the student to do.

If a student is using any type of magnification device, the teacher may still need to enlarge the original graphic, lest there be minute details the student must explore. In this situation, the TSVI would need to ensure that the student has the skills to survey an image globally before fixating on one area.

In addition, it is important to pay attention to how the graphic becomes enlarged. At the risk of losing image resolution, the teacher should not simply enlarge the graphic on a copy machine. Students with low vision can usually use computerized programs to ensure that they experience the same image resolution as do their peers, but the TSVI must take care of the type of file they are enlarging, as some image files are vector based (such as .EPS files); these file types maintain their resolution better than raster file types (.JPG, .PNG, etc.).

Other possibilities include projecting images on a whiteboard or other large screen for exploration or using a larger computer screen that can display the enlarged image. Projecting larger images is not only beneficial for students with low vision; it allows all students present to access the same image, thus promoting a more inclusive classroom for all students.

Accommodation 2: Conversion to Text

For any student with visual impairments, there are situations when the graphic can provide the same information when converted to text. If, for example, an assignment presents the results of a poll asking people which flavor of ice cream they like the most as a bar graph and asks students to answer questions related to the data (i.e., "Out of the people polled, how many people like vanilla?"), then providing students with a data table or writing out the data in prose can achieve the goal. Captioning graphics or describing images using alt-text are other common examples of this accommodation, but alt-text is not appropriate for many graphics within mathematics due to their complexity.

TSVIs should bear in mind that conversion to text is not always an option: If a mathematics lesson's purpose is to teach students to read and interpret a bar graph, for example, then converting the graphic to text would undermine the lesson. If the aforementioned ice cream poll lesson above was intended for young students to learn the parts of a bar graph, how to acquire data from the table, and how to use the bar graph to compare information, then conversion to text would provide the students with the answers—but they would do it without developing conceptual understanding.

Accommodation 3: Sonification

Multiple existing and emerging accommodations provide students with visual impairments with access to graphical information by utilizing a combination of audio information, audio cues (sonification), and even haptic vibration, in the case of some tablet devices. For example, the Orion TI-84 Talking Graphing Calculator by Orbit Research (orbitresearch.com/product/orion-ti-84-plus) utilizes direct audio output, vibration, and sonification to allow students to better understand the graphs. This device combines vibration, volume control, and stereo sound to replicate tracing the graphics. Similarly, the Desmos Graphing Calculator software (desmos.com/calculator) uses a combination of sounds to signify the left-to-right tracing motion of the graphic.

Another example of the use of data sonification is available on the NASA website (nasa.gov/data-sonifications). NASA has provided numerous data points that teachers can easily turn into lessons to explore the sonification of data with students. This data can serve as a basis for lessons that explore statistics, mean, median, and mode, as well as trends in data. As these accommodations are currently under development, they are not currently mainstream accommodations. As technology advances, there may come a time when students with visual impairments receive the choice of how to receive graphical information, even using these novel approaches.

Accommodation 4: Tactile Graphics

For blind students (those who cannot access print), the primary accommodation for access to graphical information is the use of *tactile graphics*. Simply put, tactile graphics are "graphics intended to be read principally by touch rather than vision" (Aldrich, Sheppard, & Hindle, 2003, p. 284). Tactile graphics (or *tangible graphics*) made with a variety of materials and production methods all use *relief* (changes in depth) to represent points, lines, textured patterns, and areas.

While tactile graphics have been the standard accommodation for students with severe visual impairments for years, there are still some unique challenges related to their use that TSVIs should keep in mind:

- Graphics to be converted are often "complex and cosmetically elaborate and may include three-dimensional depth" (Smith, 2017, p. 488) and thus difficult to make accessible.
- Tactile graphics can be quite difficult to create in a usable form. While use of some basic technology can enable the enlargement of a print graphic, the creation of tactile graphics may require specialized technologies and materials, as well as unique expertise, to create them. While textbook transcription providers may have individuals with specialized training available to create tactile graphics for textbooks, the responsibility of creating classroom-level tactile graphics typically rests upon the TSVI or other professional working with the students (orientation and mobility specialist, transcriber, paraeducator, etc.).
- Math educators rely upon the ability to create graphics often not found in the textbooks during their instruction and may have an expectation that they can create visual aids "on the fly" on the board while they teach. Thus, the time it takes to create tactile graphics is a constant challenge, and TSVIs and mathematics educators need to collaborate to ensure that students who need accommodations have access to them.
- Finally, school districts often do not have adequate access to the materials, software, and/or technologies required to create high-quality tactile graphics.

Even with these challenges, it is imperative that TSVIs work diligently to provide students with access to graphical information, particularly in the mathematics classroom. The purpose of this chapter is to provide TSVIs with an overview of strategies to not only create tactile graphics, but to also teach students to read, interpret, and even draw graphics.

UNDERSTANDING TACTILE PERCEPTION

Vision is a powerful sense for many reasons. Vision enables a person to simultaneously perceive all parts of an object—its totality along with its relationship to other objects in space. Vision quickly provides complex information on the qualities of size, shape, color, depth, and location. Any type of visual impairment affects the entire process of information gathering (Klatzky & Lederman, 2011). For example, the individual with visual impairments must rely on "sequential observations," which means they can only perceive part of a graphic or object at one time and thus must "build" the entire image from part to whole. This can be as true for individuals with low vision as for those who are functionally blind, depending upon their individual type of vision loss.

Another challenge created by vision loss is the ability to understand the relationship of objects to one another. Skills negatively impacted by any vision loss include depth perception, perspective, and understanding of object position (Klatzky & Lederman, 2011). The level of mental effort required to combine partial, sequential information into a single image is greater than that needed for concept formation through immediate visual perception (Klatzky & Lederman, 2011). The TSVIs or other education professionals who create tactile graphics aligned with the mathematics curriculum will need to keep these elements of tactical cognition in mind as they build useful accommodations for students with visual impairments who need graphical information.

CONSIDERATIONS FOR CREATING CLASSROOM TACTILE GRAPHICS

In thinking about creating graphics for students with visual impairments, it is important that we separate graphics into two subcategories based upon their purpose. First, we have graphics created by professional content creators for specialized textbooks (such as those created in large print and braille). The professionals who create these types of graphics, who often have unique backgrounds in areas such as graphic design or art, must adhere to professional standards. For tactile graphics, professional producers utilize the Braille Authority of North America (BANA) *Guidelines and Standards for Tactile Graphics*

2022 (2022; brailleauthority.org/guidelines-and-standards-tactile-graphics). This guidebook provides in-depth guidance on how to plan and create high-quality tactile graphics, as well as many useful examples for vision professionals.

However, most TSVIs who create tactile graphics for students within the classroom do not have the time, expertise, materials, or tools to create the level of tactile graphics found in textbooks that follow the BANA specifications. That does not mean that vision professionals cannot learn from the specifications as they build tactile graphics for their students; rather, they should use them as a resource that provides excellent guidance and examples. The next few paragraphs summarize some of the key ideas of the guidebook that professionals who work directly with students should know.

Key Idea 1: Determining When a Student Needs a Tactile Graphic

BANA's first key idea is determining whether the creation of a tactile graphic is necessary. When an educator provides a vision professional with a graphic, no one should assume that it needs a tactile version. For example, if the text repeats information already provided by the graphics, then the graphic is not necessary. Presley and Hasty (2005) created a superb "Decision Tree" (Figure 8.1) to use when determining whether a tactile graphic is necessary.

Key Idea 2: Tactile Graphics Are Not Exact Reproductions

If the TSVI determines that a tactile graphic is necessary, they must keep in mind that "a tactile graphic is a representation of a print graphic designed in a manner that is the most meaningful to the reader. It is not an exact reproduction" (BANA, 2022, p. 2–1). This is critically important to understand as many TSVIs attempt to create an exact replication of the print graphic. The complexity of some graphics may challenge the TSVI's own knowledge and expertise, and the result may be such a complex graphic that their students may not be able to use it.

As time is typically one of the greatest challenges associated with creating tactile graphics, educators should plan ahead to determine what information should be on the graphic. Students may not always require an *exact* replica of the print graphic, so Presley and Hasty (2005) recommend the TSVI begin by asking these basic questions when considering the creation of a tactile graphic:

Figure 8.1 Tactile Graphics Decision Tree

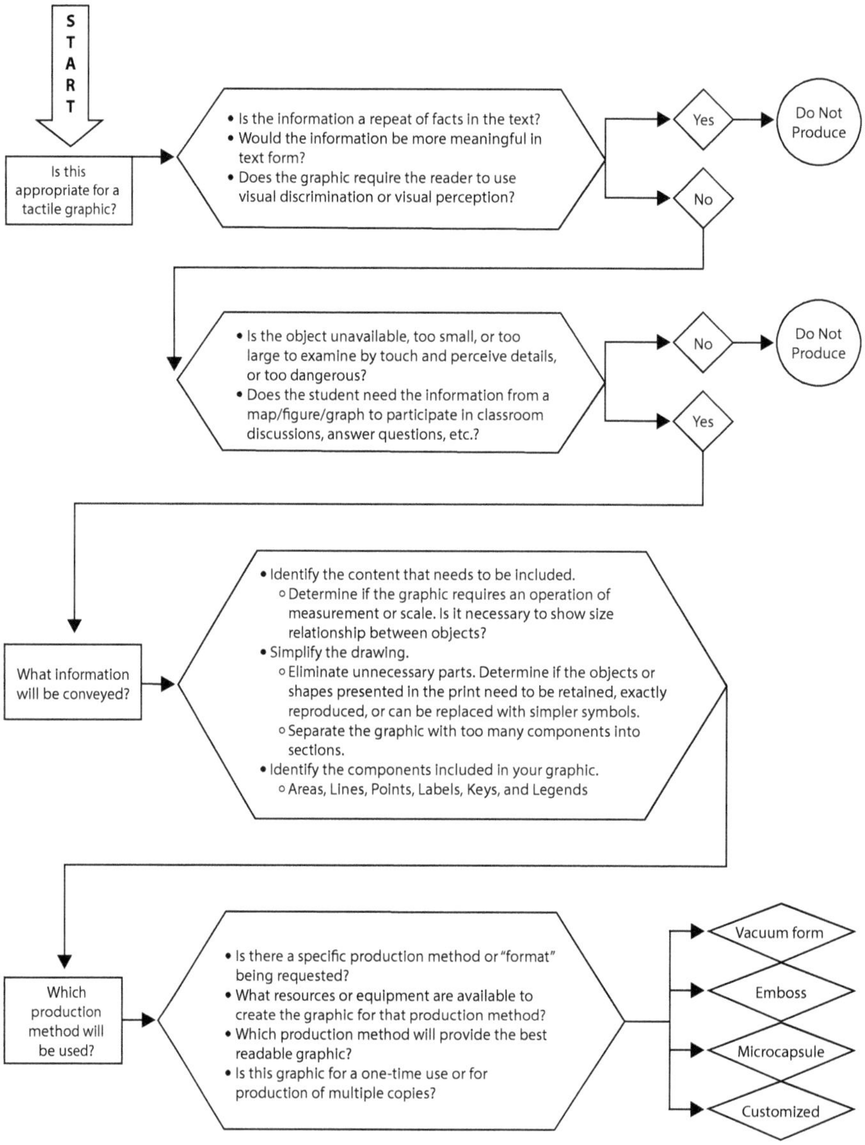

The above decision tree is a helpful workflow in deciding when and how to make a tactile graphic.

Source: Presley and Hasty (2005)

1. **Would the information be more meaningful in text form, or will the tactile graphic require a supplemental text description?** Often, tactile graphics readers require some information about the graphic before and while they explore it for context. Collaboration with the mathematics teacher is necessary to determine what type of information will help the reader understand the graphic in alignment with the lesson content. Remember, if the purpose of the graphic is to help the reader gather information, then the descriptions should not provide that information.

2. **What information must the graphic convey?**

3. **What parts must the graphic include?** This question implies (with good reason) that some information in the graphic may require simplification or removal if there is no clear purpose for its inclusion in the tactile graphic. Visually complex graphics can result in the creation of tactually cluttered graphics that may be unreadable without simplification. TSVIs must keep in mind that some information in a graphic may not be necessary for a given assignment but is still there to test the student's content understanding. The TSVI should work with the mathematics teacher to determine the essential parts of the graphic and discover those they can simplify or remove altogether.

4. **Does measurement matter, or can the TSVI resize the graphic?** As with enlargement, if the size or proportions in a graphic are meaningful to the assignment (e.g., the graphic is of an object that the student will measure using a ruler), resizing it would change the answer to the assignment. However, if size does not matter, then teachers are free to enlarge the images as needed.

5. **Does the position of objects in relation to other objects in the graphic matter?** Some tactile graphics may require rearrangement of its parts to make it less complex for the tactile reader. For example, graph titles often appear in a text box inside a bar graph or line graph. The simple act of removing that header and placing it above the graph makes it less complex for the tactile reader.

6. **Can the TSVI simplify the graphic without losing its meaning?**

7. **Are there parts of the graphic the TSVI can remove without changing the meaning of the graphic?** Again, it is important to collaborate with the math teacher to determine which parts of the graphic are unnecessary for learning purposes.

8. **Is it crucial to replicate the exact shapes in the graphic, or can the shapes appear in a simpler form?** For example, consider a pictograph that uses fish to represent the number of fish caught by three individuals. Would using a circle or square instead of a fish suffice in that graph? (This type of modification is a broadly accepted practice endorsed by the BANA *Guidelines and Standards for Tactile Graphics*, 2022.)

9. **Can the TSVI divide the graphic into separate graphics if it is too complex as one tactile graphic?** This is typically not ideal, but a student may need to explore the complexity of the graphic. For example, if multiple bars with differing shading and colors appear for each individual data point on the x-axis of a bar graph, the graphic may be too cluttered to interpret. Thus, it may be appropriate to create multiple bar graphs to represent the different bars of each data point on the x-axis.

10. **How will the student identify the parts of the graphic?** Will the student use braille labels, a key, or a legend to identify components? Labels should appear near the part of that graphic requiring a name. However, TSVIs should place the labels on the graphic in a systematic way that does not clutter the image but serves to provide additional information.

Key Idea 3: Basic Principles for Creating Tactile Graphics

The Canadian Braille Authority (CBA) and BANA formed a joint committee with the goal of creating standardized guidelines for creating tactile graphics. As part of this work, they commissioned the Graphic Research and Standards Project, or GRASP (CBA, 2010), which studied the aspects of design that provided for the best discrimination of the parts in a tactile graphic. BANA incorporated the findings of that study into their *Guidelines and Standards for Tactile Graphics 2022* (2022). The committee established the following principles from the GRASP study, along with concepts outlined within the *Guidelines and Standards for Tactile Graphics*, rooted in the idea that every graphic consists of parts that are either points, lines, area, labels, or textures (filled areas).

Points

The first and most basic element within tactile graphics is *points*. Points in tactile graphics can be any type of shape; however, the square, triangle, and circle are the easiest to recognize, while more complex shapes, such as the cross and star, may be more difficult to recognize. The minimum size of a point symbol should be $\frac{1}{4}$ inch (6.35 mm) or greater. Empty space around the points greatly improves the readability of the symbols. Within textured areas (which we will discuss later), the easiest shapes to identify are the square, circle, and triangle. They also are the easiest to read on an untextured background or on a finely dotted background.

Lines

Lines are the next attribute of the tactile graphic. Lines are easiest to follow when the lines are bold, solid, dashed, or dotted, and a thin dashed line is easier to follow than a thin solid line. Lines with irregular shapes (e.g., zigzag, dash–dot–dot–dash) are more difficult to follow. Often, lines will have symbols on the line. In those situations, hollow-shape symbols are easier to read than solid-shape symbols, and adding empty space around the point increases the user's ability to discern their shape. When lines intersect, lines of the same strength (such as bold) are difficult to distinguish from one another. Different line types (such as solid versus dashed) are easier to follow when they cross each other.

Within mathematics, many situations require grid lines, particularly when creating statistical graphs (such as bar graphs) and algebraic graphs (e.g., lines, parabolas, etc.). A grid against a light-colored field provides the best background for a reader attempting to read the bolder plotted lines or bars of the graph. When creating statistical graphs (such as bar graphs), it may be best to only create grid lines perpendicular to the bars. For example, if the bars are vertical, then horizontal grid lines will provide students with the information they need to determine what the bars represent. When creating lines within a grid, it is critical to use bold, solid lines that are significantly different from the grid lines.

Areas

Areas are larger parts of a graphic typically enclosed by straight lines, curved lines, or combinations of the two. When creating areas, variation in height between areas helps users recognize the differences. Two areas beside each other must have distinct textures or fill and feature a distinct line to separate them.

Labels

Labels, found in many print graphics, help with understanding; however, TSVIs may need to add them to a tactile graphic to aid conceptual understanding. They should place labels in the most appropriate location that may or may not correspond to the placement of the labels in the print graphic. When placing the label, it should be between $\frac{1}{2}$ inch and $\frac{1}{4}$ inch (3.18 mm and 6.35 mm) from the component to which it belongs. Otherwise, the TSVI should add lines connecting the label to the item it defines and may make the tactile graphic more complex. The use of empty zones around a label increases the reader's ability to discriminate it from the background texture.

Textures

Textures function within areas, but the texture applied provides the reader of tactile graphics with more information. Textures can include "solid fill," "large dot pattern," "diagonal lines," "small-dot pattern," and "smooth/no fill." The purpose of the textures is to aid the reader in determining differences between areas next to each other. For example, a map may use one texture for the land while it uses a different texture for water. In mathematics, one texture may be used to show the shaded portion of one inequality while another texture is used for showing the shaded portion for a separate inequality.

PRODUCTION PROCESSES FOR EDUCATIONAL PROFESSIONALS

There are myriad ways to create accessible graphics, each with its own pros and cons. For TSVIs, the chosen production processes will ultimately depend upon the available tools and materials combined with the amount of time available to create the graphic.

For creating enlarged print graphics for students with low vision, the two most common options are preprinting enlargement (working with digital files) or printing enlargement (enlarging a hard copy of the graphic). Preprinting enlargement requires either having access to the digital file of the graphic (such as a .JPEG, .PNG, or .EPS file) or scanning the print graphic for editing. The positive of this method is that it often retains the colors of the graphic and works across multiple platforms. The challenge is that it requires specialized software and access to graphics that withstand editing without a loss of resolution. TSVIs should use vector-based image files (e.g., .SVG, .EPS) whenever possible instead of raster images (e.g., .JPEG, .TIFF, .PNG) when resizing an image.

Printing enlargement is often a simpler method, as it typically only requires the print graphic and either a scanner or copier. Many copy machines have built-in software that allows for enlarging (or shrinking) an image, and some machines can print on larger paper (such as 11 × 17 paper, also known as A3). If the machine allows for color printing, it can maintain the color attributes while replicating the image. The challenge of this method is that it may require copying parts of the image and then combining them (such as taping the picture together). Another challenge is that many copiers that are available to educators do not print in color, thus the images may need to be black and white instead. Depending on the image and the purposes of color, this may be an acceptable accommodation. Of course, students with low vision may benefit more from applicable assistive technologies that magnify the original image instead.

The production of tactile graphics is ultimately dependent upon the tools, materials, time, and ability of the TSVI. Professional tactile graphics producers use a variety of methods to produce tactile graphics that are often beyond the scope of the TSVI. The methods below are some of the most common methods that TSVIs use to produce "quick" tactile graphics for the classroom.

Vacuum Form

Within most braille textbooks, students will have access to tactile graphics in the form of vacuum form sheets (e.g., Brailon thermoform paper, available from americanthermoform.com/brailon-thermoform-paper). More commonly referred to as *thermoform pages*, these are sheets of plastic heated and melted on top of a mold of the tactile graphics.

If a TSVI has access to a vacuum form machine, they can create the tactile graphics using a variety of materials. For example, the TSVI can create a tactile graphic using a *collage method* with a variety of materials, such as drywall tape, string, felt paper, etc. Another common way to create the template mold is to manipulate metal diagramming foil sheets (such as those available from the American Printing House for the Blind, aph.org/product/aluminum-diagramming-foil-sheets) using a variety of hand tools that create the tactile aspects. With either of these methods, the vacuum form sheet, placed atop the mold, melts over it to form the tactile graphics. This is an inexpensive method of mass-producing images, as the mold is reusable, but it does require some skill to create the initial mold.

Figure 8.2 Geometric Shapes on Thermoform Paper

Thermoform paper can create sharp contrast for tactile graphics.

Fuser

Another common means of creating quick tactile graphics is to use a fuser and microcapsule paper (Figure 8.3). Using this method, the TSVI can add the graphic to the microcapsule paper using a photocopier, printer, or certain types of pens, then run the paper through a special fuser machine that heats the paper and raises the gray and black areas of the paper to form the tactile graphic. This method is very efficient, particularly for black-and-white graphics that need crisp angles (such as in geometry). However, it does require specialized equipment and materials, and the microcapsule paper can be expensive.

Figure 8.3 Creating Tactile Graphics With Capsulated Paper

A TSVI shows a teacher how to create a tactile graphic using capsulated paper running through a fuser machine.
Source: Tiffany Wild

Braille Embosser

Primarily, braille embossers quickly "print" braille documents by forming braille dots on braille paper. However, many newer braille embossers have the capability to create tactile graphics using the dots it can form for braille, similar to a dot-matrix printer making an image from dots only. Braille embossers often require special software to create embossable graphics. The benefit of this method is that users can save the graphics as a file to inexpensively emboss as needed. The two key challenges are that it requires access to a braille embosser that can create tactile graphics, and users need design skill using specialized software. There are some braille embossers that are capable of embossing at different heights to provide depth to the tactile graphics.

Specialized Tactile Creation Products and Technologies

Recent years have seen the development of new types of paper, such as tactile drawing film. This film expands when pressed down and "puffs up" to create a tactile image of the pressed area. APH products such as the Draftsman Tactile Drawing Board (aph.org/product/draftsman-tactile-drawing-board) and the TactileDoodle (aph.org/product/tactiledoodle-kit), which provide a tablet that can hold the film and provide a rubber mat beneath, commonly use this paper. Tactile drawing film provides instantaneous tactile graphics and is relatively affordable. However, there is a limited level of granularity, and thus thin lines can be difficult to create.

Another APH product is its Quick-Draw Paper (aph.org/product/quick-draw-paper), which allows the user to draw on the paper using a water-based marker; the water makes the paper rise to create a tactile image. This paper is quite good for creating quick tactile graphics within the classroom, but it can be expensive for some schools (though it is eligible for purchase using Federal Quota funds).

Sidebar 8.1 Students Creating Their Own Accessible Graphics

While the primary focus of this chapter is on education professionals creating accessible graphics for students with visual impairments, one must not overlook the fact that students will need to create their own graphics.

Strategies for Creating Graphics for Students With Low Vision

For students with visual impairments who are able to read and write print, the type and level of accommodations for drawing their own graphics will depend on their functional vision. For example, for many students with low vision, using a pencil may not provide a dark enough line for them to easily see the line. Therefore, erasable pens or thin-lined markers may be crucial for drawing assignments. Using specific tools for drawing—such as a ruler, protractor, or compass—also requires consideration. With all of

Sidebar 8.1 *(continued)*

these items, students will need to have either accessible tools (discussed in Chapter 3) or instruction in how to adapt their low-vision-friendly tools to complete their drawings.

Strategies for Creating Graphics for Students Without Vision

For students without functional vision, creating graphics will depend on the goal of the activity and the materials needed to create the graphics. For example, if the goal of the lesson is to create a representation of a linear function (a line), then the students could use strategies discussed in Chapter 6 to create the line using various materials on a tactile grid (paper or board). Creating graphics with the use of tactile objects (such as Wikki Stix, strings, other materials, etc.) is akin to the "collage" technique and may be appropriate for certain tasks.

Beyond the collage technique, students will need to rely on specialized paper that creates raised lines. Students could use APH's Draftsman, TactileDoodle, or Quick-Draw Paper to create their own tactile graphics. While students are able to "free draw" on these papers, it is more useful to use tools such as rulers, compasses, and tracing templates to draw more exact shapes. With direct instruction on how to use these tools, students with visual impairments can construct many of the lines and shapes used in geometry and algebra.

One of the burgeoning areas for tactile graphics is the development of multiline braille displays that can render text and/or graphics. This technology is essentially based upon the "dot matrix" concept in which the braille dots create an image. Currently, there are three devices on the market that create tactile graphics using braille dots.

The Monarch (by HumanWare and APH; aph.org/meet-monarch) uses 2,880 individual dots and creates both 32-cell braille text and static graphics through its 10-line, 96 × 40 display and 3,840 pins. Users can utilize the Monarch as a standalone device to read and write braille while also outputting tactile graphics.

Orbit Research's Graphiti (orbitresearch.com/product/graphiti) uses 2,400 dots to create static tactile graphics. This device allows for variation of the pins (which create the dots), and it features the option for users to draw graphics on the device by touch using a finger or using its own software. It produces a true "dot matrix"-style output to create images or braille, although the dots do not provide braille according to braille specifications.

Finally, Tactile Engineering's Cadence Tablet (tactile-engineering.com/cadence) uses a modular system of smaller devices, with each device having 384 dots. Users can connect up to four separate devices to create lines of 24 cells each that can produce braille and static and dynamic (moving) tactile graphics.

As this technology has advanced, the ability to create tactile graphics using one of these devices may provide a more efficient means of creating classroom-level tactile graphics.

In-Class Adaptations

These four production methods are the most common ones a TSVI may utilize. However, other methods for creating "quick" tactile graphics provide a basic level of access when time is the greatest challenge. For example, the creation of simple geometry shapes is very common in elementary classrooms and even in geometry and/or algebra in high school.

A simple technique to create quick graphics requires braille paper (11 × 11 in., or 27.94 × 27.94 cm, is best), a rubber mat, a pencil, possibly a ruler, and a tracing wheel used in marking patterns when sewing. To create a simple graphic, simply place the braille paper on the rubber mat and draw the shape using the pencil. Next, use the tracing wheel to trace the lines. When the user flips the paper over, they should find the shape embossed as raised lines. This application can help users create basic shapes, draw simple lines (such as two parallel lines and a transversal in geometry), and construct many other applications. The only caveat is that the images will be backward when the user turns the paper over, so more complex shapes will require careful planning. APH offers a Tactile Graphics Kit (aph.org/product/tactile-graphics-kit) with which users can create more complex graphics using this method.

An array of other options exists for creating tactile graphics. For example, TSVIs can employ the collage method mentioned earlier to create an original graphic that the student will use directly instead of creating a mold first. Many teachers use materials that they can easily acquire at craft stores, such as string, felt paper, buttons, Wikki Stix (wikkistix.com), etc., to create tactile graphics within the classroom. The benefit of creating collage tactile graphics is that users can create them using common materials, but the downside is that they lack durability suitable for heavy use. Another option is to use commercially available products specifically made for creating tactile graphics. Examples include the Tactile Book Builder Kit (aph.org/product/tactile-book-builder-kit), Tactile Town: 3-D O&M Kit (aph.org/product/tactile-town-3-d-om-kit), Tactile Connections kits (aph.org/product/tactile-connections-symbols-for-communication), and the Picture Maker Wheatley Tactile Diagramming Kit (aph.org/product/picture-maker-wheatley-tactile-diagramming-kit). All these kits are available from APH and use a collage method that features felt or plastic materials and a Velcro backing to affix the graphics to a felt board.

The particular production techniques will depend upon the available resources, complexity of the graphic, and time needed to create it before giving it to the student. Using tools such as the collage kits, providing tactile drawing film or Quick-Draw Paper, or using braille paper with tracing wheels are all techniques that other educators, paraeducators, or even other students can learn for quick creation of tactile graphics within the classroom. While the TSVI may want to spend more time creating more sophisticated tactile graphics that students can use multiple times, these less complex strategies should be sufficient to provide student access to mathematical graphics as needed within the classroom.

TEACHING GRAPHIC LITERACY

While ensuring that students with visual impairments have access to graphical information is critical, there is a separate aspect of access that TSVIs often overlook. The challenges of access to graphics and the strategies for creating accessible graphics are the most often discussed topics in the literature and in training of vision professionals. But even when the students with visual impairments have access to high-quality graphics, they still require direct instruction on how to "read" the graphics. In the same way that we focus on literacy and numeracy, students with visual impairments need instruction in graphical literacy (known as *graphicacy*) (Aldrich, Sheppard, & Hindle, 2003).

In general, all students with visual impairments (low vision or blind) need training in strategies on how to read a graphic to glean the information needed to learn the concept under investigation. For students with low vision, the primary instruction by the vision professional should focus on them using either their residual vision or appropriate assistive technologies in a systematic way. Students with low vision have a particular breadth of acuity and field of vision, and so instruction will require an individualized approach. While the strategy for teaching a student with low vision to read graphics may vary, here is an example of an approach:

1. Orient the student to the way the graphic lies in front of them. Be sure the graphic faces the proper way and is not upside down or sideways. Note any orientation markers on the graphic.

2. As the student explores the tactile graphic, have them verbally say what they find in the graphic on each of the following steps.

3. Visually scan the graphic to find the larger parts of the graphic. Look for labels or other important features.

4. Read the title of the graphic, if provided. Read the labels and connect each to its corresponding part of the graphic. Read any legend or key provided.

5. Scan the graphic in a systematic way to find the minute parts of the graphic. For example, start in the top left-hand corner and move to the right until the graphic ends. Finally, shift down and read to the left, making a back-and-forth sweep of the graphic, looking for the details of the graphic.

6. For any details that are difficult to read, remind the student to ask for more information or use assistive technology tools to read those smaller issues. Often, small details are important in mathematical graphics, so students need instruction on looking for detailed information.

During instruction, the TSVI can provide the student with a variety of mathematical graphics. The key is to ask them to explain all the details they encounter in order for the TSVI to ensure the student can read and interpret all of the diagram's important graphic representations. This is a very important step as the TSVI teaches graphical literacy. Once the student has mastered a strategy, the student can repeat it as they encounter more graphics in the math classroom.

For blind students, the instruction must be more purposeful, as they need to develop an array of skills to develop tactile graphic literacy. As noted earlier, touch (1) requires the ability to read graphics using part-to-whole strategies; (2) requires more time; (3) does not allow for instant recognition, especially from different perspectives; (4) does not easily discriminate scale; (5) does not easily discriminate spatial relationships; and (6) relies heavily upon labels and supporting information for identification and context. Therefore, students do not learn to read tactile graphics in one lesson but instead through a series of experiences that take place throughout their academic career.

Exploration

There are numerous educational frameworks that promote a sequence of representation that shifts from concrete to pictorial to symbolic; these steps can help to maximize mental or abstract concept development. Early tactile experiences must focus on touching "real" (i.e., concrete) objects before developing concepts of representations. During this early exploration, students should investigate basic concepts such as shape, hardness/softness, texture, and relative size and weight. Young students need opportunities to explore objects, follow their contours, examine various textures, and begin the processes of tactile discrimination. It is appropriate to make connections between real objects (a concrete object such as a ball) and a pictorial representation of a sphere (a picture of a ball) before moving to a symbolic representation (a tactile graphic of a ball).

Texture Discrimination

Students will need to learn to discriminate among different types of textures such as smooth, rough, and "bumpy" (with different levels of granularity) quite early in their education. TSVIs should provide students with opportunities to explore different types of materials and learn to differentiate between the textures using either common items (plastic sheets, felt, etc.) or products such as APH's Flip-Over Concept Books: TEXTURES (aph.org/product/flip-over-concept-books-textures).

Shape Identification

Through exploration, students need to develop the concepts of basic shapes and be able to identify them early. This includes being able to differentiate among a point, a line, and an area. In their exploration of areas, students will apply their texture discrimination skills as they learn to distinguish among different types of patterns across different areas. They should also have opportunities to connect two-dimensional shapes (such as a circle or square) to their corresponding three-dimensional shape (e.g., sphere or cube). While exploring the shapes, they should learn to discriminate between and identify the parts of the shapes, including the edges (formed by lines or curves) and any points on the line (such as at the intersection of two lines). Students must have opportunities to learn how to identify shapes—whether they take up an entire page or are small enough to feel under a finger—as basic shapes often represent items on tactile graphics.

Lines

Points, lines, and areas form graphics, but lines are the most critical aspect of any graphic. The same is true for any tactile graphic, as the lines (straight or curved) provide the general shape of the objects in the graphic. Therefore, students will need to have ample opportunities to learn how to follow lines with their fingers. TSVIs should provide students with worksheets that have various types of lines (solid bold, solid thin, dashed, zigzag, double dashed) that also move in different directions (vertical, horizontal, diagonal, etc.). The ability to trace a line and then describe it will help students develop skills needed for graphical concept development. The TSVI can create these worksheets or introduce a tool such as the APH Flip-Over Concept Books: LINE PATHS (aph.org/product/flip-over-concept-books-line-paths).

Spatial Understanding

One of the challenges created by vision loss is its impact on spatial understanding (Jones, 1975). Students with visual impairments must learn how to explore graphics and understand the relationships among objects. It is even more challenging when depth or perspective exists within the graphic, as there is not enough information provided for most readers to determine position.

While using tools such as APH's Tactile Town or StackUps Kit (aph.org/product/stackups-kit-spatial-reasoning-using-cubes-and-isometric-drawings) help with these concepts, their use of concrete, three-dimensional materials to teach the concepts can be limiting. Despite having a great deal of practice, an advanced tactile graphics reader still may not truly grasp the perspectives within the graphic and thus may require an explanation. In mathematics, this becomes a major issue when students are studying three-dimensional shapes, surface area, and volume.

Scale

Students with visual impairments often have difficulties understanding scale (Jones, Taylor, & Broadwell, 2009). Trying to understand the scale used to represent objects that are very large or very small in tactile graphics can be difficult, especially if students cannot directly experience the object. It is important to explicitly teach the scale representations of any images to ensure conceptual understanding of the object represented in the graphic.

Systematic Scanning

All the previous tactile graphics literacy skills should constantly include a focus on teaching systematic scanning. There is no right or wrong way to read a tactile graphic, but a systematic approach tends to yield better results. The TSVI should teach the students multiple scanning strategies but ultimately allow them to determine which method works best for them. Similar to the system used by low-vision students, here is one approach:

1. Some tactile graphics may include a transcriber's note or description of the graphic on a separate page. The students should begin by reading this information before reviewing the graphic for context. When given a tactile graphic, the students should first do a quick scan of the entire document to establish an overview. Some TSVIs recommend using a vertical search pattern for this initial scan, as it allows the students to cover more area quickly with fewer scanning sweeps. Some students tend to place both their hands on the entire graphic and then fixate on anything that seems unique. However, students instead need to learn

to quickly scan the tactile graphic and then immediately look for a title or header.

2. The students then should look for any other information provided, such as a legend, key, or braille labels.

3. At this point, the students will need to start systematically reviewing the graphic. Based upon their individual preferences and abilities, students may use both hands or one hand. Students will most likely use a horizontal pattern, as this is the most familiar pattern they use to read print or braille. However, a student may decide to use a vertical pattern or even a circular pattern to find all the parts of the graphic.

4. Once the students have scanned the entire graphic, they are now ready to conduct a more focused review of the graphic. One common strategy is for each student to place one hand on one part of the graphic and establish it as the "anchor" while they explore the rest of the graphic with their other hand. This allows the students to systematically review the graphic, including the parts of the tactile graphic, as they mentally create the whole image.

Regardless of the strategy the students use to gather information from a tactile graphic, they will most always need to have some extra information provided, either in the text or with an oral description, to understand the graphic's context and parts. The TSVI must focus on purposefully providing students with opportunities to explore many different types of tactile graphics and have them practice explaining what they are learning from the graphic.

CONCLUSION

Students encounter graphics throughout mathematics; their exposure begins with basic shapes, then moves on to advanced geometric figures and statistical graphics, then to algebra and calculus graphics to represent functions. The ingrained use of graphics within mathematics instruction makes teaching effectively without them nearly impossible. Therefore, it is imperative that TSVIs work diligently to not only make certain that the students have access to high-quality tactile graphics but also to develop a sequence of lessons that ensure the students develop strong graphic literacy skills.

REFERENCES

Aldrich. F., Sheppard, L., & Hindle. Y. (2003). First steps towards a model of tactile graphicacy. *The Cartographic Journal, 40*(3), 283–287. https://doi.org/10.1177/026461960202000203

Braille Authority of North America (BANA). (2022). *Guidelines and standards for tactile graphics 2022.*

Canadian Braille Authority. (2003). *Report of tactile graphics sub-committee part 3.* https://www.brailleliteracycanada.ca/storage/standards/Report_Tactile_Graphics_part3.pdf

Jones, B. (1975). Spatial perception in the blind. *British Journal of Psychology, 66*(4), 461–472. https://doi.org/10.1111/j.2044-8295.1975.tb01481.x

Jones, G., Taylor, A., & Broadwell, B. (2009). Concepts of scale held by students with visual impairments. *Journal of Research in Science Teaching, 46,* 506–509. https://doi.org/10.1002/tea.20277

Klatzky, R. L., & Lederman, S. J. (2011). Haptic object perception: Spatial dimensionality and relation to vision. *Philosophical Transactions of The Royal Society, 366,* 3097–3105. https://doi.org/10.1098/rstb.2011.0153.

Presley, I., & Hasty, L. (2005). *Techniques for creating and instructing with tactile graphics.* American Foundation for the Blind.

Smith, D. W. (2017). Mathematics. In C. Holbrook, T. Wright, & C. Kamei-Hannan (Eds.), *Foundations of education* (3rd ed., Vol. II, Instructional strategies for teaching children and youths with visual impairments, p. 488). AFB Press.

CHAPTER 9

MATHEMATICAL SYNERGY: UNLOCKING THE POTENTIAL OF TEACHER COLLABORATION

Amanda Lannan

INTRODUCTION

The U.S. Bureau of Labor Statistics (2023) published a projection of the 10 fastest growing occupations in the next decade. All 10 career fields identified—including wind turbine technician, physician's assistant, and software developer—require a broad range of knowledge and skills in science, technology, engineering, and mathematics (STEM). Occupational projections and other current trends indicate that opportunities and learning in STEM will heavily influence the career trajectory of today's students, and STEM skills will apply to almost any profession.

Students with disabilities are as likely to enter STEM fields as those without disabilities (Sutton, 2017). However, the results of the 2019 Programme for International Student Assessment (PISA) revealed stagnant and declining performance in mathematics for many students on the margin (Organisation for Economic Co-operation and Development [OECD], 2023), including students who are blind or visually impaired. Addressing the mathematics performance gap is critical for the future success of these students in the workforce.

Facing Obstacles

There is a widening gap between the math achievement of students who are blind or visually impaired and nondisabled students (Bell & Silverman, 2019; Cryer, 2013). An examination of studies on mathematical proficiency among students who were blind or visually impaired revealed that 75% were at least one grade behind their nondisabled peers, and 20% were four or more grades behind (Gulley et al., 2017). Students who are blind or visually impaired frequently identify the visual aspects of mathematics as the primary barrier they face. Although the visual components of math instruction do present challenges, there are many more barriers preventing these students from accessing the math curriculum, including a lack of accommodations for students, inaccessible curricular materials, and limited knowledge by educators about how to teach students who are blind and visually impaired.

It is crucial to develop a solid grasp of number sense before moving on to more advanced mathematics (see Chapter 2, "Foundations of Numbers and Operations"). There is a documented link between number sense and visual processes (Anobile et al., 2013; Dakin et al., 2011; Stoianov & Zorzi, 2012; Zhou et al., 2015); however, evidence shows that individuals who are blind or visually impaired may not rely on visual processes to develop this understanding (Ahlberg & Csocsán, 1999; Rottmann et al., 2020). Research by Crollen and Collignon (2020) demonstrates that people who are blind and visually impaired engage regions of the brain responsible for working memory, tactile manipulation, and language for these tasks instead of engaging visually. Students who are blind and visually impaired still perform well in number sense tasks, but they often understand numbers through words rather than tactile exploration (Ahlberg and Csocsán, 1999). The research shows that there is substantial evidence that vision is not necessary for the development of number sense or other math skills and concepts, and that students who are blind and visually impaired can develop these skills using nonvisual pathways (Ahlberg & Csocsán, 1999; Amalric et al., 2018; Crollen & Collignon, 2020).

Barriers to Mathematical Success

In the face of obstacles posed by my own vision loss, my unyielding curiosity and fierce determination to maintain independence have remained unwavering. My tenacious parents and teachers provided the real-life references and materials I needed to achieve academically. Equally important was that these open-minded, supportive, and caring teachers and specialists believed I should have the same educational opportunities as my peers. The general population has limited interactions with individuals who have a visual impairment; as a result, they may hold common preconceptions, attitudes, and beliefs surrounding visual impairments. Despite legislative assurances of "equal opportunity," the struggle to actualize this promise persists, and the need to continuously seek solutions by which to access otherwise inaccessible information is an ongoing endeavor. The greatest challenge lies in reassuring others that, with accommodations in place and creative problem solving, individuals with visual impairments can perform at a level equal to that of their sighted peers.

Students with disabilities encounter numerous accessibility challenges throughout their educational journey, particularly in math and science. For students who are blind and visually impaired, the inherently visual nature of math and science is often the access barrier that comes to mind (Bell & Silverman, 2019). Although the visual aspects of math and science instruction do create challenges for blind and visually impaired students, there are many other barriers that limit their access to learning math: Lack of appropriate accommodations, inaccessible curricular materials, and educators with limited expertise in teaching students with visual impairments are the overarching barriers (Bell & Silverman, 2019). Closely examining these access barriers presents undeniable evidence of how a lack of equitable access impedes students who are blind and visually impaired from fully engaging in mathematics learning.

Throughout their education, students who are blind and visually impaired repeatedly face a predominant reliance on visual instruction in STEM disciplines. Teachers often use visual techniques such as writing on the board and then referencing specific excerpts using phrases such as "Record the answer here" or "Look at this figure" while teaching mathematical and scientific concepts (Jitngernmadan et al., 2017). These techniques obviously limit access for students who are blind or visually impaired; they also may negatively affect students' perceptions about their capacity for taking STEM courses.

Spatial reasoning can also be a challenge for individuals who are blind or visually impaired due to limited instruction in methods of exploring and interpreting the tactile graphics necessary for comprehension of certain math and science-related concepts (Smith & Smothers, 2012). Students with little to no vision rely on meaningful instruction—including verbal explanations, descriptions, and real-world activities—to develop an understanding of math concepts that sighted students learn incidentally by observing everyday activities (Turnbull et al., 2013). For example, sighted children observe patterns in their surroundings, such as stripes on a shirt, tiles on the floor, or alternating colors in a sequence. They learn to recognize and create patterns, which strengthens their logical thinking skills. They develop a visual understanding of money by observing the everyday use of bills and coins. However, students who are blind or visually impaired rely on explicit instruction and tactile representations. Taking the idea of incidental learning one step further, traditional methods of teaching begin by looking at the big picture and then breaking the concept down into parts. Sighted individuals can look at a car and see the wheels, windows, bumpers, etc. On the other hand, students who are blind or visually impaired explore each part separately and use what they have learned to construct the "big picture" of the car. A student relying on nonvisual methods to learn math, rather than visually scanning graphics and spatial arrangements, may experience a heavier cognitive load, making it more difficult for them to process information as quickly as do their peers (Jitngernmadan et al., 2017).

Do Perceptions Matter?

Productive student-teacher relationships positively impact equity in mathematics classrooms (Bartell, 2011). Although the intention is not to communicate low expectations to students, the reality is that implicit biases and stereotypes based on disability, socioeconomic status, and racial and/or ethnic identity do exist. These negative perceptions can impact learning and achievement, as well as student attitudes and motivations (Bayram et al., 2015; Brophy & Good, 1974). Bartell (2011) discovered that positive student-teacher relationships connect to positive feedback and genuine interest in students' mathematical thinking. When caring and committed educators communicate their authentic understanding of their students and their students' lived social realities, equitable learning opportunities are more likely (Boaler & Staples, 2008). As both a blind student and an educator, I have firsthand experience in how teacher expectations play a critical role in student achievement.

When a teacher learns that a student with visual impairments will be in their math class, the all-too-frequent response is disbelief, resentment, or sheer panic. Given the low-incidence nature of blindness and visual impairments, mathematics educators often do not feel equipped to provide students with visual impairments with effective instruction (Ajuwon et al., 2015; Maguvhe, 2015; Rule et al., 2011). Likewise, when teachers of students with visual impairments (TSVIs) may dread working with students in higher level STEM courses, as they often feel that they lack the necessary content knowledge and confidence in their ability to convey the unfamiliar material correctly (Hawley et al., 2013; Herzberg & Rosenblum, 2014; Supalo et al., 2014). The student may unintentionally interpret these feelings of inadequacy as their own inability to successfully learn math (Supalo et al., 2014). Unfortunately, the misconceptions surrounding disabilities and mathematics often put students receiving special education services at a serious disadvantage due to a lack of access to the material and, potentially, teacher disposition (Lambert, 2015).

To evaluate the distribution of meaningful learning opportunities among all students, Reinholz and Shah (2018) offer three basic questions related to equity:

- Who participates?
- What is the nature of that student's participation?
- What opportunities do teachers make available to support that participation?

They found that educator expectancy bias contributes to perceptions that students with disabilities, as a consequence of their disability, cannot learn from or engage in discussion-based learning (Lambert, 2015; Lambert & Tan, 2017; Reinholz & Shah, 2018). These deficit perspectives are more evident in the sequential progression of increasingly advanced courses requiring students to successfully complete prerequisites prior to promotion. The opportunity for a student to progress to the next level is contingent on their ability to meet or exceed current expectations. This linear process can directly limit the opportunities students with disabilities have to participate in higher level mathematics (Jackson & Neel, 2006; Kurz et al., 2014). Consequently, students with disabilities have very limited power to make decisions regarding their education, such as placement in advanced mathematics courses. The current state of discriminatory practices reduces not only opportunities for students with disabilities to access equitable academic instruction but also significantly limits these students' access to postsecondary choices and career paths. Many people perceive a "visually impaired" student as within a problem category defined by all the perceived deficiencies associated with having a visual impairment. People often overlook their assets, learning strengths, lived experiences, and unique ways of thinking and reasoning mathematically (Tan et al., 2019). Purposeful and focused actions necessary to harness the collective skills of educators will ensure that all students receive the education they deserve (Friend & Cook, 2020).

In inclusive mathematics classrooms—especially those serving students who are blind and visually impaired—paraprofessionals play a vital role in ensuring students have access to, participate in, and benefit from instruction.

Paraprofessionals' backgrounds range from having extensive classroom experience to being new to education. Their perceptions and understandings of visual impairments can vary widely due to differences in training and prior experience. Some paraprofessionals may hold misconceptions about blindness, leading them to offer well-intended but counterproductive support—such as hovering too closely or completing tasks for the student—thereby limiting the student's independence and social interaction (Tannock, 2009). In some cases, they may also lack the specialized knowledge needed to support math learning, including familiarity with Nemeth code, tactile graphics, or accessible digital tools. To address these challenges, targeted professional development is crucial. Opportunities such as attending conferences, reading specialized publications, and participating in collaborative training can help paraprofessionals develop more accurate understandings and implement effective support strategies (Holbrook, 2015).

Parents and families play a pivotal role in shaping the academic outcomes of blind and visually impaired students, particularly in STEM subjects. Because children spend more time with their families than with any other educational partner, the attitudes and expectations held at home often have the greatest influence. Research consistently shows that parent behaviors and beliefs significantly affect children's participation, performance, and dispositions in STEM subjects (Ray et al., 2009). When families support high expectations and foster independence, they lay the groundwork for academic success and resilience in challenging fields such as math and science.

However, not all parental influences are equally constructive. Some families may hold negative or limiting beliefs about disability, which can unintentionally restrict a child's access to learning experiences and tools, such as braille, tactile graphics, or assistive technology (Koenig & Holbrook, 2000; Palawat & May, 2012). Overprotectiveness, fear, or lack of information can result in fewer opportunities for students to engage with rigorous content or develop problem-solving skills essential for STEM learning. In contrast, when families view blindness as one component of their child's identity and encourage self-advocacy and exploration, students are more likely to develop confidence and persistence in academic pursuits (Castellano, 2005; Ferguson, 2002).

These findings suggest that improving access to highly qualified educators and appropriate learning resources is essential for supporting math achievement among students who are blind and visually impaired. It is urgent that all students have access to the educational resources necessary to reach their highest potential without barriers, either physical or perceived. It is equally imperative that mathematics teachers, TSVIs, paraprofessionals, and families approach this challenge in the spirit of collaboration, taking each student's individual experience and capabilities into consideration.

PEDAGOGICAL APPROACHES

General education mathematics and special education mathematics take distinctly different approaches in their philosophies, methods, and even curriculum design (Little & Dieker, 2015). Instructional pedagogy, from a special education perspective, provides students with strategies, explicit instruction, and repetition to develop procedural competencies. Special education teachers purposefully design learning opportunities to include the students' accommodations and academic needs (Johnson et al., 2019), with the goal of providing access to the general education standards. Students in special education receive instructional placement across a continuum of services, in the least restrictive environment, most appropriate to each student (IDEA, 2004). Because of the widespread attitude that general education students are capable, have an innate conceptual understanding, and possess excellent problem-solving skills (Tan et al., 2018), they more often receive instruction through an inquiry-based approach that builds on their conceptual knowledge (Akyuz & Stephan, 2020).

General education mathematics teachers and special education educators typically take different approaches to teaching. In one case, when students with disabilities participate in mathematics education programs with nondisabled peers, a general education teacher who has developed strong mathematical and pedagogical content knowledge provides the content instruction (Pearl et al., 2012; Scruggs et al., 2007). Alternately, when students receive special education services in mathematics *outside* the mainstream setting from a specialist with extensive preparation to meet the educational and legal nuances of individual education programs (IEPs), that specialist may not have sufficient expertise in mathematics instruction. Many special education teachers work across grade levels and mathematical concepts, and it is difficult to be a content expert at every grade level.

In today's schools, the majority of students with disabilities receive instruction in general education mathematics classrooms from a general education teacher, who is the content expert, and support from the special education teachers to provide the students' accommodations and meet academic needs (Johnson et al., 2019). Therefore, synergy within educational teams has become critically important. Simply placing students with visual impairments in inclusive settings does not ensure equitable opportunities for them to engage in high levels of mathematics. To provide academic instruction and ensure students with visual impairments can access the knowledge and skills often gained visually, a collaborative approach is essential.

UNIVERSAL DESIGN FOR LEARNING

The concept of Universal Design for Learning (UDL) stems from the universal design principle of architecture in which structures and buildings are accessible for all by design (Burgstahler, 2015; Rose & Meyer, 2008). The foundations of UDL encourage the same approach to curriculum, such as in mathematics. The UDL approach focuses on "equity and access" for all students by embracing learner variability through (1) multiple ways to engage students in learning, (2) multiple ways to present content, and (3) multiple ways for students to express what they are learning (Wehmeyer, 2006). This set of principles is critical in all settings but particularly essential for students with disabilities. Most applications of UDL principles provide for considered variability in representation, engagement, and expression for students with a range of academic or motivational needs by removing barriers in the curriculum.

Accessibility is not beneficial without ensuring usability. Students must have the training and support to build competencies. Even with a strong UDL approach to instruction, students need support from teachers who understand how to adapt instruction to meet their needs.

Sidebar 9.1 Implementing the UDL Framework

Implementing the UDL framework to design inclusive mathematics lessons can seem overwhelming. While building lessons, the initial challenge is to design them with UDL in mind. In a classroom with multiple students, it can be daunting to design a lesson plan for a concept that allows the learner to present, engage with, and express the content. However, if math lessons are built upon UDL principles, then not only will the students with visual impairments benefit, but there is great potential for all students to benefit as well. To help teachers get started, Novak Education

Sidebar 9.1 *(continued)*

offers tips to implement the UDL guidelines with a specific math focus. These tips (found at https://info.novakeducation.com/udl-guidelines-math) can help educators provide the scaffolding and accommodations needed to improve outcomes for all students. In addition, elementary, middle, and high school "lesson plan makeovers" (novakeducation.com/blog/udl-lesson-plan-makeovers-math-lessons-math-resource-download) demonstrate ways of creating lessons that embrace learner variability, target grade level standards, and foster expert student learning. The UDL principles and practices serve as a guide to support the entire educational team with instructional designs that do not require students to engage, represent, or express themselves identically.

Co-Teaching

Co-teaching is an instructional model designed to serve students with disabilities in the least restrictive environment. In co-teaching, researchers (e.g., Salend, 2011; Villa & Thousand, 2005) recognize that inclusive practices offer benefits to students with disabilities; however, not all agree that the general education setting is appropriate for students with specific disabilities (Solis et al., 2012). A survey of 498 early childhood teachers revealed a significant dichotomy in teachers' opinions on who should receive inclusive education. Less than half of respondents felt inclusion was appropriate for students identified as having intellectual disability, physical disability, visual impairment, hearing loss, autism spectrum disorder, or attention-deficit hyperactivity disorder (Lee et al., 2015). Despite this disparity in opinion, schools commonly practice co-teaching to support students with disabilities in the general education classroom (Friend et al., 2010). In this model, a general education teacher and a special education teacher work as a team to coplan, co-instruct, and co-assess (Bauwens et al., 1989; Cook & Friend, 1991;1995; Dieker & Murawski, 2003; Magiera et al., 2005; Scruggs et al., 2007).

The way math educators and TSVIs might collaborate to support students will vary, depending on the needs of the particular student, the curriculum design, and the instructional approach. When describing the distinct teacher roles in the coplanning phase, the mathematics teacher typically focuses on ensuring access to concepts, peer dialogue, and conceptual understanding, all while maintaining the pace and trajectory of mathematical thinking. The

TSVI provides specialized instruction in the expanded core curriculum, such as the compensatory skills needed for using the abacus or talking calculator, sensory efficiency skills to improve tactile discrimination for learning how to systematically interpret the tactile graphs or charts, and if needed, reading and writing math notation using braille. TSVIs also prepare adapted materials in advance of the lesson so as to involve their students in activities at the same time, not later than the rest of the group. These proactive measures facilitate the students' ability to understand and reproduce the mathematical relationships in the format that best meets each student's individual needs.

The individual roles of the general education teacher and TSVI should complement each other. As the content expert, the mathematics teacher uses their collective experiences and knowledge to direct student learning. TSVIs focus on skill development and access to the instructional components. TSVIs have expertise in identifying and mitigating the educational impact of a visual impairment and the comorbidity of other conditions. Nevertheless, it is important to empower both teachers to use best practices in mathematics instruction as well as in special education. Combining these disciplines through co-teaching collectively can help teachers meet students where they are; at the same time, they can support one another to ensure equitable access to both the context and content of the lesson.

A TEAM APPROACH

Collaboration goes beyond the math educator and TSVI to include all members of the educational team. Effective collaboration involves utilizing the strengths and expertise of all team members. TSVIs bring specialized knowledge of accessible tools, tactile graphics, manipulatives, and assistive technology. Orientation and mobility (O&M) instructors contribute insights into how students navigate space and materials, which is essential when teaching geometric and spatial concepts. General educators provide the curricular framework and context for instruction. Meanwhile, parents offer valuable information about their child's learning behaviors and needs in non-school settings. Legislation such as the Individuals with Disabilities Education Act (IDEA) have facilitated the transition to more inclusive educational practices. As more students with disabilities join the general education classroom, the need for all educators to acquire more knowledge to meet the needs of every learner has become even more important.

One way for teachers to gain more information about their students with disabilities is to collaborate with the specialists and service providers who interact with these students every day. Speech and language pathologists (SLPs), O&M specialists, physical therapists, occupational therapists, and counselors know and understand their students—but only within their own settings. This is why collaboration among team members is crucial. What children learn during each part of their day must transfer consistently throughout their entire day. It is not productive or beneficial for knowledge and skills to be applicable only in one setting. Students must be able to use their skills in all areas of learning and in life. Collaboration could expand teacher knowledge so that students receive the needed accommodations and supports across all subjects. For example, in a math class, suggestions from an SLP could improve the methods of communication, input from an assistive technology specialist might offer solutions to access needs, and the occupational or physical therapist may provide options for an ergonomic workstation that allows for a proper viewing position to avoid back and neck strain.

Far too often, there is a presumption that service providers and specialists may not have the background to contribute to mathematics instruction, or mathematics teachers may assume these other professionals and experts do not have a role in their content. However, all parties listed above add a clear dimension of UDL to the classroom and create pathways to learning that go beyond auditory and visual input. It is essential that all members feel equally able to contribute to the collaborative team's common goals and establish a shared expectation among team members that blind students can and should meet the same rigorous academic standards as their sighted peers (Castellano, 2005). Despite differences in experience and skills, a common understanding helps the team achieve its shared goals. As collaborative teams can have diverse members, establishing each member's role and contribution is foundational. Maintaining continuity depends on consistent goals and unified messages among members. Embracing the help and knowledge provided by others will better equip teachers to create the best learning environment possible.

Distributed Leadership

Distributed leadership is a theoretical framework that underscores the significance of shared decision making and cooperation among various education stakeholders, including school personnel, the student, and the student's family.

The framework advocates for leadership that transcends an individual's role, cultivating a culture of collaboration and collective responsibility (Harris, 2019). To be successful, all members of the team must view themselves as partners in the educational process. A common thread across definitions of teams is that teamwork is a process among partners who work together to achieve mutual goals. Teamwork allows individuals—and, as a result, the team—to grow professionally. It is necessary to solicit input from all team members, ask and answer necessary questions, and share ideas. Teamwork does not happen by accident. It requires effort, commitment, and a willingness to accept and work through the challenges of collaboration.

BUILDING BLOCKS OF COLLABORATION

In a recent literature review, Griffiths et al. (2021) identified core components of successful collaboration: relationship building, shared values, active engagement, and collaboration. Using these key elements, they developed a comprehensive definition and model of collaboration.

Figure 9.1 Building Blocks of Collaboration

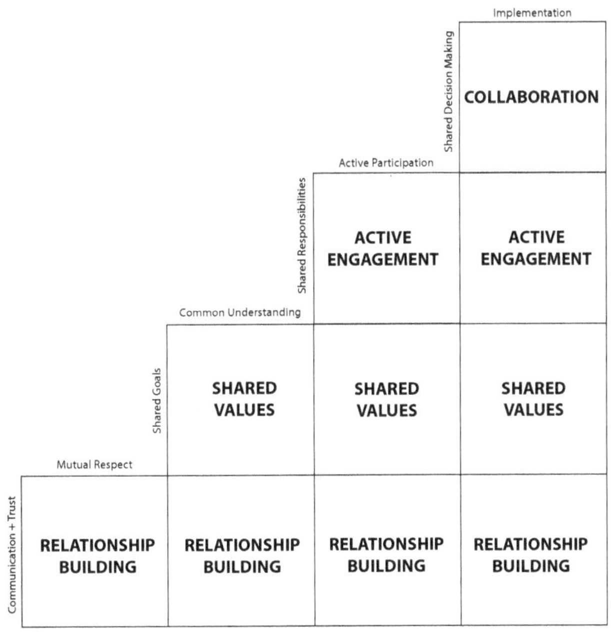

Source: Griffiths et al. (2021)

Although personal characteristics such as openness to new ideas are important to the process, Griffiths et al. assert that how team members leverage individual strengths for collective success are the fundamental building blocks of effective collaboration.

The "building blocks" model developed by Griffiths et al. (Figure 9.1) portrays collaborative development as an ongoing, multifaceted process in which each step toward collaboration is built upon and reinforced by the previous step. Active, two-way dialogue throughout all phases is vital for teams to provide student-focused mathematics instruction. Team members support one another as they strengthen their collaborative methods at each phase of the model before moving on to the next level.

Relationship Building

Communication, trust, and mutual respect are the fundamental elements of building a strong, collaborative relationship within and across teams (Griffiths et al., 2021). At this initial stage, an emphasis on these elements is critical to the overall integrity of the team.

Communication

The key to maximizing the impact of teacher collaboration is effective communication. The team's goal should be the same: to maximize teacher impact in order to meet every student's needs. One of the best ways for team members to cultivate a positive relationship is to get to know one another; building in social opportunities is key, especially in the early stages of a team's collaboration. Making time to have lunch, get coffee, or have a drink after school—whatever works best for the team—will give the team members the opportunity to get to know one another. Learning about one another's lives helps team members see one another as people, not just fellow educators. From there, members can work together to iron out the details by sharing strengths, challenges, passions, and working and communication styles. As a team begins to work together, members should find ways to incorporate everyone's strengths so that students will benefit from them.

Another important element of team communication is scheduling time to check in. Team members should ask, "How do you feel like it's going?" and then genuinely listen to their colleagues' responses. When issues arise, team members should ask whether there is anything they can do to help and provide suggestions if appropriate.

Finally, each team member must hold themself to the same expectations they have for others. If team members want their colleagues to have a growth mindset and hold themselves accountable, they must do the same, admitting when they make a mistake and following the same approach when attempting to rectify a situation. Equally important is offering specific, thoughtful, genuine compliments. Teaching is tough work, but reminders of what they are doing well make team members feel supported and ultimately do a better job.

One of the biggest challenges co-teachers face is how they can maintain an interpersonal relationship with peers (Hourcade & Bauwens, 2003). There is no right way to communicate. However, when teachers respect their teaching partner's communication–collaboration style, they enhance the likelihood of sustaining an effective partnership. Without open communication or compromise, effective communication is not possible.

Open communication involves active listening. In the case of a math educator and a TSVI, it is important each recognizes the other as an equal, credentialed teacher with their own area of expertise. Clearly defining roles and responsibilities in the partnership highlights the strengths they offer each other. Strong communication also involves open lines of communication in both formal and informal situations. Accepting members' diverse backgrounds and being open to different viewpoints fosters productive dialogue. The need for effective communication extends beyond the TSVI and math teacher: Interactions with paraeducators, related service providers, and the student and their family can offer valuable ideas and resources.

Including Student and Family Voices

Collaboration with families is a cornerstone of effective educational practices for students with disabilities. Family-centered practices emphasize the importance of respecting and integrating the perspectives of parents and guardians. This means striving to understand their lived experiences, cultural values, emotional responses, and long-term hopes for their children. As emphasized by Lalvani (2012), families have the right to participate in virtually every aspect of their child's education. Teachers who respect this right by interacting with empathy and professionalism can build strong partnerships that benefit students academically, socially, and emotionally (Cianca & Wischnowski, 2012; Francis et al., 2016).

Family involvement begins with listening. Parents of children with disabilities often navigate complex emotions, including grief, ambivalence, and optimism (Douglas, 2014; Ferguson, 2002). Some parents may grieve the loss of expected life experiences for their child, while others may feel ambivalence due to the difficult decisions they face. Still others, particularly those with strong support systems or a positive orientation toward disability, may respond with hope and determination. Understanding these diverse reactions—shaped by culture, personal experiences, and access to resources—is essential for meaningful collaboration (Palawat & May, 2012; Ray et al., 2009).

Educators must also recognize that cultural and linguistic differences can affect family–school communication. Parents from diverse backgrounds or who have a history of difficult experiences in their own education may experience schools as intimidating or unfamiliar environments. People from other cultures in particular may hold different beliefs about disability or be unsure of the expectations for their involvement. Using interpreters effectively, learning key phrases in the family's home language, and asking positive, inclusive questions can promote comfort and trust (Cummings & Hardin, 2017; Olivos, 2009).

In addition to involving families, it is critical to empower students with disabilities to take an active role in decisions about their own education. *Self-determination*—learning to express preferences, set goals, and advocate for oneself—is a vital skill that students can foster from an early age (Hart & Brehm, 2013). While the level of student involvement varies depending on age and individual ability, educators should provide opportunities for all students to participate in developmentally appropriate ways. For example, elementary students might describe what they enjoy learning, middle school students might explain their learning needs and goals, and high school students may take the lead in their IEP meetings (Griffin, 2011; Nolan-Spohn, 2016).

Approaches such as *student-led IEPs* and *person-centered planning* offer structured ways to support student voice and vision. These strategies help students take ownership of their educational experiences, increase their confidence, and build critical life skills (Test & Neale, 2004; Wehmeyer et al., 2013). Planning tools such as MAPs, PATH, and Circle of Friends (see Sidebar 9.2) provide visual, collaborative formats for defining a student's future aspirations and the steps needed to achieve them.

Authentically including families and students in educational planning results in decisions that are more likely to reflect the students' strengths, preferences, and long-term goals. The consistency created between home and school supports reinforces learning and ensures a better position for students to succeed both academically and in life beyond school. Ultimately, honoring the voices of both families and students fosters a culture of respect, equity, and shared responsibility.

Trust

Team members develop trust over time when they invest in accessible communication systems and take the time to understand that not everyone communicates in the same ways. With trust comes greater vulnerability and openness. When members recognize that the team pursues shared goals, behaviors are consistent and reliable, and freely exchanged ideas lack judgment, trust thrives. Days are long and the work can be overwhelming, but taking time to check in with one another about how each person is doing fosters relationships and boosts cohesion among the team.

Sidebar 9.2 Three Powerful Tools for Person-Centered Planning

The following tools share a common goal: empowering individuals by centering their voice, building a team around them, and turning their dreams into achievable plans.

MAPs (Making Action Plans)

MAPs is a structured, person-centered planning process designed to help individuals—especially those with disabilities—identify their hopes for the future and develop a step-by-step plan to reach them. A team that includes the individual, family, friends, and professionals works together to understand the person's life story, dreams, concerns, and strengths. Through open dialogue, the team creates specific action steps, defines roles, and builds a support system that empowers the individual to move forward with confidence.

PATH (Planning Alternative Tomorrows With Hope)

PATH is a dynamic and visual planning tool that begins with a compelling vision of the individual's "dream future" and works backward to identify short- and long-term steps needed to reach that vision. The process is typically facilitated by trained leaders and uses colorful charts or drawings to map out goals, resources, timelines, and people who can help. PATH

Sidebar 9.2 *(continued)*

helps individuals stay motivated and focused while encouraging collaboration and accountability from their support network.

Circle of Friends

Circle of Friends focuses on building meaningful social connections and reducing isolation for individuals with disabilities. A small group of peers meets regularly to offer friendship, support, and inclusion in school or community life. While not a formal planning tool like MAPs or PATH, Circle of Friends plays a vital role in person-centered approaches by fostering relationships, strengthening communication skills, and helping individuals feel valued and connected.

Mutual Respect

Developing mutual respect involves openly sharing ideas and a willingness to be flexible. Disagreements are inevitable, but respecting one another is imperative. Flexibility allows team members to gain another person's perspective on how to possibly improve instruction, what methods might be best for assessment, or which resources to use. Mutual respect requires open-mindedness and a willingness to try something different.

Mutual respect emerges as members value one another's competencies and flexibility. Regardless of an individual's role in the education of students, upholding mutual respect requires understanding one another as people; each team member's role in student support; and how each team member empowers everyone else's collective awareness, understanding, and knowledge of student needs. Reciprocal communication among team members is essential to maximize educational access in the least restrictive environment. Teams must continually nurture open communication, trust, and respect throughout collaboration.

Shared Values

When a team has begun the process of developing a strong relationship, members can then identify and use their areas of expertise to mutually determine a set of shared goals and objectives (Cowan et al., 2004).

Shared Goals

At the core of collaboration is a shared goal or a common problem that brings the team together to complete a task that they otherwise cannot accomplish independently (Fewster-Thuente, 2015; Olson, 2003). To develop shared goals, each member of the team provides input, and the team mutually creates and agrees upon goals.

Sidebar 9.3 Mutual Respect in Action

In a traditional middle school math class, students use their notes to quiz one another for an upcoming geometry quiz. However, there is a student in the class who is blind and reads braille. The TSVI suggests a tactile, quiz-like review. At first, the math teacher is skeptical about the time and materials needed to prepare, but after some discussion, they agree to try the TSVI's suggestion. The teacher divides the review into concept stations. After collecting the items, the two instructors create task cards for each of the review questions. Some examples include using popsicle sticks to create different types of triangles; assembling polygons with LEGO pieces that match a specified number of sides and vertices to demonstrate key terms such as parallel, perpendicular, and intersection; and applying student knowledge of angles, rotation, and symmetry through the art of origami. All students, including the student who is blind, actively engage with the activity and even come up with new and challenging questions. Two days later, the scores reflect that the review was a success. The thrilled teachers agree to use the activity next year.

When a math teacher and a TSVI work together, all students can benefit from the rich expertise and experiences of both educators. General educators tend to have a broad knowledge of the curriculum, standards, and desired outcomes for the larger group, while TSVIs have expertise in individualizing curriculum and instruction to meet the needs of students with blindness and visual impairments. Both viewpoints hold significance, and the team should collectively address them. All teachers—general education and special educators alike—are responsible for ensuring student success. A single shared goal transforms these separate common goals for student success into a single shared goal, ultimately cultivating trust.

It is crucial that each team member feels equally empowered to make contributions toward the shared objectives of the collaborative team, as emphasized by Griffiths et al. (2021) and Hallam et al. (2015). The intention is to establish interdependence, where individual expertise is necessary to accomplish the team's overall tasks.

Sidebar 9.4 A Collaborative Lesson Plan Example

Following the principles of UDL, the TSVI and math teacher brainstorm a lesson about fractions.

Objective: Students will be able to measure ingredients using fractional parts as part of a multisensory, inclusive, and authentic learning activity.

1. **Introduction (engagement)**: Begin by discussing the importance of fractions in everyday life, such as in cooking recipes, measurements, and sharing equally. Show a video about cooking with audio description. Ask students to share what they know about fractions.
2. **Presenting the content (representation)**: Introduce the concept of fractions using a variety of methods:
   a. **Visual:** Use visual aids such as fraction bars or pie charts to represent fractions and their parts.
   b. **Auditory:** Explain the concept verbally, providing clear and concise descriptions.
   c. **Tactile:** Provide comparable tactile materials, such as textured fraction circles, magnetic fraction bars, or a fraction tower.
3. **Hands-on activity (engagement)**: Engage the students in a hands-on activity where they explore fractions. This could involve:
   a. **Fraction manipulatives:** Provide tactile fraction manipulatives, such as high contrast and braille-labeled fraction cards and tactile models (such as those found in APH's MathBuilders Fractions kit, aph.org/product/mathbuilders-fractions).
   b. **Collaborative group work:** Encourage students to work in groups or play fraction games, such as those found on the We Are Teachers website, such as Connect Fractions (played with a repurposed Connect 4 set) or Fraction War (which uses a basic deck of cards); both activities can be found at weareteachers.com/fourth-grade-math-games. Encourage students to share their findings and tactile experiences.

Sidebar 9.4 *(continued)*

4. **Practice and application (expression):** Give students a set of fraction problems to solve. Provide worksheets in accessible formats such as large print and/or braille, or in an accessible digital format. Allow students to choose how they want to represent their answers: drawing, in writing, via a video of them using manipulatives to solve, or by creating tactile representations.
5. **Summarization and discussion (engagement/representation):** Conclude the lesson by summarizing key concepts and discussing how people may use fractions in various real-life situations. Encourage students to share their findings and experiences.
6. **Assessment (expression):** Assess students' understanding through various means, such as traditional written assessments, oral presentations, or tactile representations. Ensure accommodations are in place for students to demonstrate their knowledge effectively.

Common Understanding

Students who are blind or visually impaired benefit from the diverse experiences, cultures, and expertise contributed by the individuals within their educational team. Developing continuity in collaborative relationships relies on the common understanding that the team cannot accomplish their responsibilities without everyone working together. A fundamental element of the process is developing common messages and understanding among team members (Glueck & Reschly, 2014). The team's success depends on each person's ability to coordinate their efforts with others. Interdependent teams listen, respect others, and embrace their responsibility to the group.

While collaborating with professional colleagues such as general education teachers, special educators, SLPs, or school psychologists is important, it is equally important to establish productive partnerships with paraprofessionals.

Understanding the Paraprofessional's Role in Math Instruction

Paraprofessionals (also known as paraeducators or instructional aides) are school staff members who assist certified teachers in implementing educational programs. When supporting students who are blind or visually impaired in mathematics, paraprofessionals may help with tasks such as:

- preparing and adapting tactile or large-print materials

- supporting students with math manipulatives or assistive technology
- providing one-on-one reinforcement of math concepts introduced by the teacher
- documenting progress or observations during math lessons

Unlike with relationships with professional colleagues, a teacher's partnership with paraprofessionals involves an element of supervision: assigning tasks, clarifying instructional strategies, and orienting them to classroom procedures and expectations specific to supporting students who are blind or visually impaired. This supervisory dynamic requires clear communication and purposeful collaboration.

Clarifying Expectations and Building a Working Relationship

It is essential to clarify responsibilities early. At the beginning of the year or when a new paraprofessional joins the classroom, take time to:

- provide a physical space for their materials
- explain classroom routines and behavior expectations
- identify specific ways they will support students during math instruction
- encourage them to ask questions or raise concerns

Frequent early communication helps prevent confusion, especially as paraprofessionals may serve multiple classrooms with varying instructional styles. Consider holding initial planning meetings that include the general education teacher, TSVI, and paraprofessional to align expectations and avoid miscommunication.

Collaborative Practices for Math Instruction

Despite the supervisory–subordinate nature of the relationship, paraprofessionals can be active, collaborative partners in supporting mathematics learning. Effective collaboration may include:

- joint problem solving around student learning needs
- planning adaptations to math lessons or manipulatives
- contributing insights based on close observation of a blind or visually impaired student's performance

Importantly, paraprofessionals should reinforce content already introduced by a certified teacher. No one should ask them to deliver initial instruction or make instructional decisions independently. When introducing math concepts for the first time—especially concepts that involve abstract or spatial reasoning—certified professionals should lead, while paraprofessionals should follow up with practice or clarification.

Teachers should communicate openly to paraprofessionals about what they are doing well and where adjustments may be necessary. Positive feedback and respectful redirection build trust and ensure consistency in supporting student learning.

Sidebar 9.5 Collaborating to Support Maya in Middle School Math: An Example

Background

Maya is a seventh-grade student who is blind and uses a combination of braille, Nemeth code, and a refreshable braille display to access academic content. She receives support in her general education math class from two teachers: Ms. Johnson, the math teacher, and Mr. Reyes, a paraprofessional assigned to assist her during core instruction periods.

Initial Challenges

At the beginning of the school year, Ms. Johnson and Mr. Reyes experienced confusion about their roles. Ms. Johnson was not sure what Mr. Reyes could or should do during lessons, and Mr. Reyes often waited passively for direction. Maya sometimes missed opportunities to participate in class because of delays in receiving tactile materials or unclear instructions.

Collaborative Problem Solving

During a team planning meeting facilitated by Ms. Paris, Maya's TSVI, the team identified the following action steps:

- **Defined roles:** Ms. Johnson would lead all initial instruction and introduce new math concepts, while Mr. Reyes would reinforce content through guided practice using braille materials.
- **Material preparation:** The TSVI and Mr. Reyes would collaborate to ensure braille versions of worksheets and manipulatives were ready in advance. Mr. Reyes would receive these materials at least 1 day before each math lesson.

Sidebar 9.5 *(continued)*

- **Check-ins:** Ms. Johnson and Mr. Reyes scheduled weekly 5-minute check-ins to clarify upcoming activities, debrief subsequent challenges, and celebrate successes.
- **Student independence:** Mr. Reyes received informal coaching from Ms. Paris on strategies to promote independence, such as verbal prompting instead of hand-under-hand assistance and encouraging Maya to advocate for her needs.

Outcomes

Over the following months, Maya's participation in math improved significantly. She was able to follow lessons more independently, complete problem-solving tasks using Nemeth code more confidently, and contribute to group discussions frequently. Ms. Johnson reported feeling more confident in guiding Mr. Reyes, and Mr. Reyes expressed that he felt respected and valued as part of the instructional team.

This case demonstrates how intentional communication, role clarification, and mutual respect can transform a paraprofessional-teacher partnership into a powerful support system for a student with visual impairment studying mathematics.

Having a common understanding does not mean that everyone on the team understands how to graph parabolas or the differences between Unified English Braille (UEB) Math/Science (complete braille code that includes math) and UEB with Nemeth (the standard braille code for representing math and science expressions embedded within the UEB context). An interdependent team with a common understanding recognizes that a blind or visually impaired student who wants to go to college for engineering will need to take advanced mathematics courses, which will require specially designed instruction to ensure the student has equitable access to a meaningful mathematics education. For the student's dream of being an engineer to come to fruition, all members of the student's educational team will need to coordinate their efforts. Regardless of the learning objective (e.g., learning how to use the accessibility features on Desmos, simplifying rational expressions, or navigating the busy high school campus), each professional plays an essential role in educating blind and visually impaired students.

Active Engagement

Fire and rescue crews work as a team in emergency situations. Medical personnel combine their knowledge and skills to complete a surgical procedure. Multiple construction crews work together to build a house. In these examples, each individual's success has a direct impact on the overall success of the team. Yet, in education, teachers often find themselves alone in their classroom, struggling to keep up with the unsurmountable responsibilities alone. Educators talk to one another about assignments, projects, and strategies to support students, but sharing information is not the same as interdependent collaboration. Without coordinated effort, teachers may repeat tasks or overlook them altogether. Miscommunication can lead to confusion or misunderstandings. Teachers might not use resources efficiently. Imagine the crisis that could occur in an emergency, during surgery, or on a construction site without a collaborative and cooperative team. In the field of education, an interdependent team is no less critical, and that requires that each team member take an active and proactive role in educating students.

Shared Responsibility

Although their individual responsibilities differ, the team works together in tandem to address the academic, social-emotional, and independent functioning needs of every student, and positive results in each domain maximize each student's overall achievement. Understanding how a student's eye condition may have an adverse effect on the acquisition of mathematics is just as necessary as knowing the learning progressions in mathematics. Recognizing how different educational elements can jointly influence outcomes reflects interdependency within the team.

Active Participation

Active participation refers to the proactive involvement and engagement of team members in the collaborative process of planning, implementing, and evaluating educational strategies and interventions for the benefit of students. At this phase of collaboration, the team moves from working in isolation and then reporting the results to sharing the work and evaluating progress together. To visualize how the interactive participatory process might occur, consider the following case.

Sidebar 9.6 Fourth-Grade Student Maura: A Case Study

Background

Maura is an 11-year-old fourth-grade student at Riverside Elementary. According to her most current eye report, Maura's diagnosis is retinopathy of prematurity. Because of an enucleation in her left eye last year, she can only see hand movements with her right eye. Maura received academic functioning assessment using the Brigance: Her word recognition was at a first-grade level, reading comprehension at an upper first-grade level, sentence writing at a second-grade level, spelling at a first-grade level, and math computations at less than a first-grade level. Maura receives speech as well as language and orientation and mobility services from an SLP and O&M specialist, respectively. The TSVI provides 1 hour per day of specialized direct instruction in reading and writing braille and instruction in Nemeth code. Maura spends the remainder of her day in the general education classroom with her peers, and she functions quite easily in most environments, identifying familiar voices easily. She performs age-appropriate tasks such as dressing, personal hygiene care, and organization of her personal items with assistance and support. She is able to locate and utilize necessary materials at school to complete her work, as well as independently follow her daily schedule.

Educational Team in Action

Mr. Hanna teaches fourth grade at Riverside Elementary. His class will be working on fractions over the next 2 weeks. Mr. Hanna usually introduces the unit by playing a fraction game on the Smartboard. Now that Maura is in his class, he realizes how many of his activities center around visuals. He has decided to create stations around the room, each with a different way to interact with fractions. He has a few ideas but could use some pointers about how to bring in more sensory approaches.

During lunch, Mr. Hanna sent a message to the other teachers and service providers who work with the students with disabilities in his class. By the end of the day, Mr. Hanna had a list of great ideas. The team agreed to meet in the morning before school so they could begin gathering materials.

Sidebar 9.6 *(continued)*

The Results of Collaboration

Here are two of the fraction stations the team designed, along with ideas for increasing access for students who are blind or visually impaired:

Listening to Fractions

Watch a simple cooking video. Collect data on the number of times each fraction appears in the video.

Accommodation: Use a piece of braille graphing paper, Velcro board, or magnetic board. Down the left side, place a tactile representation of each fraction found in the video. To tally the fractions quickly, cut Wikki Stix (wikkistix.com) into small pieces. When the student hears a fraction mentioned in the video, they can quickly place a sticky tally mark on the correct line.

Delicious Fractions

Using the provided art supplies, make a pizza. Cut out a large circle and cut it into 8 slices. Use different colored sheets of paper to add toppings to the pizza, such as circles (pepperoni), triangles (cheese), and squares (green peppers), making sure to use fractions to describe the portions of each topping. Have students share their creations. Teachers can easily vary the depth of questioning by asking for the fractional parts of toppings in relation to different amounts of pizza.

Accommodation: Use a pounce wheel or pattern tracer to create tactile lines for the student to follow as they cut out their pizza slices. Provide options of high-contrast colors or textures to help students better distinguish each topping.

Collaboration

Successfully progressing through the initial three stages of the building blocks model (relationship building, shared values, and active engagement) establishes a robust collaborative foundation. The ultimate objective in the collaborative process is to reach the final step: identifying solutions and putting them into action. As such, the final step of collaboration includes shared decision making and the effective implementation of these decisions.

Shared Decision Making

Shared decision making is a process involving methodically collecting input from all team members. Successful execution of the team's collective responsibility for gathering assessment data and anecdotes relies on the active involvement of all participants. Within a cohesive team that cultivates strong relationships, team members highly value the unique perspectives contributed by each expert, often generating ideas that may exceed what the team can put into practice (Hallam et al., 2015). Consequently, it is imperative for the team to establish and follow a clear procedure to determine the future course of action.

Implementation

The conclusive and perhaps most pivotal component of the collaborative process is effective implementation. Implementation, as defined by Dulaney (2012), encompasses the team's execution of its decisions while ensuring fidelity. Even if the team collaborates effectively and formulates a robust plan for the student, without successful implementation, there will be no improvement to the student's outcome. Consequently, planning, creating, and collectively implementing instruction to support blind and visually impaired students in achieving short- and long-term goals is undoubtably the most significant phase of collaboration (Hall & Hord, 2011).

CONCLUSION

The concept of what counts as successful mathematics learning is evolving, technology solutions are ever changing, and the pressure to prepare students with 21st-century skills that STEM careers prioritize is increasing. Throughout this chapter, we have discussed the common barriers to effective mathematics education for students who are blind or visually impaired, along with various methods for collaborative problem solving that provide scaffolded instruction and individualized support. A team approach offers educators opportunities for collaboration and professional support for one another. When teachers collaborate, they complement one another in their work to educate all students equitably. Instead of operating from a deficit orientation, they work collaboratively to remove the curricular, environmental, and social barriers. Adopting a mindset of assuming success rather than anticipating failure, teams will discover opportunities to provide students who or blind or visually impaired equitable access to mathematics at every level.

REFERENCES

Ahlberg, A., & Csocsán, E. (1999). How children who are blind experience numbers. *Journal of Visual Impairment & Blindness, 93*(9), 549–560.

Ajuwon, P. M., Sarraj, H., Griffin-Shirley, N., Lechtenberger, D., & Zhou, L. (2015). Including students who Are visually impaired in the classroom: Attitudes of preservice teachers. *Journal of Visual Impairment & Blindness, 109*(2), 131–140. https://doi.org/10.1177/0145482X1510900208

Akyuz, D., & Stephan, M. (2020). Co-teaching practices that build autonomy for students with learning disabilities in mathematics. *International Journal of Mathematical Education in Science and Technology, 53*(4), 1–28. https://doi.org/10.1080/0020739X.2020.1795286

Amalric, M., Denghien, I., & Dehaene, S. (2018). On the role of visual experience in mathematical development: Evidence from blind mathematicians. *Developmental Cognitive Neuroscience, 30*, 314–323. https://doi.org/10.1016/j.dcn.2017.09.007

Anobile, G., Stievano, P., & Burr, D. C. (2013). Visual sustained attention and numerosity sensitivity correlate with math achievement in children. *Journal of Experimental Child Psychology, 116*(2), 380–391. https://doi.org/10.1016/j.jecp.2013.06.006

Bartell, T. (2011). Caring, race, culture, and power: A research synthesis toward supporting mathematics teachers in caring with awareness. *Journal of Urban Mathematics Education, 4*(1). https://doi.org/10.21423/jume-v4i1a128

Bauwens, J., Hourcade, J., & Friend, M. (1989). Cooperative teaching: A model for general and special education integration. *Remedial and Special Education, 10*(2), 17–22.

Bayram, G. I., Corlu, M. S., Aydin E., Ortaçtepe D., & Alapala, B. (2015). An exploratory study of visually impaired students' perceptions of inclusive mathematics education. *British Journal of Visual Impairment, 33*(3), 212–219. https://doi.org/10.1177/0264619615591865

Bell, E. C., & Silverman, A. M. (2019). Access to math and science content for youth who are blind or visually impaired. *Journal of Blindness Innovation and Research, 9*(1).

Boaler, J., & Staples, M. (2008). Creating mathematical futures through an equitable teaching approach: The case of Railside School. *Teachers College Record, 110*(3), 608–645. https://doi.org/10.1177/016146810811000302

Brophy, J. E., & Good, T. L. (1974). *Teacher–student relationships: Causes and consequences.* Holt, Rinehart & Winston.

Burgstahler, S. (2015). *Equal access: Universal design of instruction.* University of Washington.

Castellano, C. (2005). *Making it work: Educating the blind/visually impaired student in the regular school.* IAP – Information Age Publishing.

Cianca, M., & Wischnowski, M. (2012). Collaborating with parents of students with disabilities. *Educational Horizons, 91*(1), 26–29.

Cook, L., & Friend, M. (1991). Principles for the practice of collaboration in schools. *Preventing School Failure, 35*(4), 6–9. https://doi.org/10.1080/1045988x.1991.9944251

Cook, L., & Friend, M. (1995). Co-teaching: Guidelines for creating effective practices. *Focus on Exceptional Children, 28*(3), 1–12. https://doi.org/10.17161/foec.v28i3.6852

Cowan, R. J., Swearer Napolitano, S. M., & Sheridan, S. M. (2004). Home–school collaboration. *Encyclopedia of Applied Psychology, 2,* 201–208.

Crollen, V., & Collignon, O. (2020). How visual is the «number sense»? Insights from the blind. *Neuroscience & Biobehavioral Reviews. 118,* 290–297. https://doi.org/10.1016/j.neubiorev.2020.07.022

Cryer, H. (2013, April 10). *Teaching STEM subjects to blind and partially sighted students: Literature review and resources* (Literature review #6). RNIM Centre for Accessible Information. https://www.rnib.org.uk/professionals/health-social-care-education-professionals/knowledge-and-research-hub/reports-and-insight/teaching-stem-subjects-to-blind-and-partially-sighted-learners

Cummings, K. P., & Hardin, B. J. (2017). Navigating disability and related services: Stories of immigrant families. *Early Child Development and Care, 187*(1), 115–127. https://doi.org/10.1080/03004430.2016.1152962

Dakin, S. C., Tibber, M. S., Greenwood, J. A., & Morgan, M. J. (2011). A common visual metric for approximate number and density. *Proceedings of the National Academy of Sciences, 108*(49), 19552–19557. https://doi.org/10.1073/pnas.1113195108

Dieker, L. A., & Murawski, W. W. (2003). Co-teaching at the secondary level: Unique issues, current trends, and suggestions for success. *High School Journal, 86*(4), 1–13. https://doi.org/10.1353/hsj.2003.0007

Dulaney, S. K. (2012). A middle school's response-to-intervention journey: Building systematic processes of facilitation, collaboration, and implementation. *NASSP Bulletin, 97*(1), 53–77. https://doi.org/10.1177/0192636512469289

Ferguson, P. M. (2002). A place in the family: An historical interpretation of research on parental reactions to having a child with a disability. *The Journal of Special Education, 36*(3), 124–131. https://doi.org/10.1177/00224669020360030201

Fewster-Thuente, L. (2015). Working together toward a common goal: A grounded theory of nurse–physician collaboration. *MEDSURG Nursing, 24*(5), 356–362.

Francis, G. L., Fuchs, E., Johnson, A. D., Gordon, S., & Grant, A. (2016). Developing parent–professional partnerships in a postsecondary education program for students with disabilities. *Psychology in the Schools, 53*(10), 1045–1056. https://doi.org/10.1002/pits.21974

Friend, M. (2020). *Interactions: Collaboration skills for school professionals* (9th ed.). Pearson.

Friend, M., Cook, L., Hurley-Chamberlain, D., & Shamberger, C. (2010). Co-teaching: An illustration of the complexity of collaboration in special education. *Journal of Educational and Psychological Consultation, 20*(1), 2–27. https://doi.org/10.1080/10474410903535380

Glueck, C. L., & Reschly, A. L. (2014). Examining congruence within school–family partnerships: Definition, importance, and current measurement approaches. *Psychology in the Schools, 51*(3), 296–315. https://doi.org/10.1002/pits.21745

Griffin, M. M. (2011). Promoting IEP participation: Effects of interventions, considerations for CLD students. *Career Development for Exceptional Individuals, 34*(3), 153–164. https://doi.org/10.1177/0885728811410561

Griffiths, A. J., Alsip, J., Hart, S. R., Round, R. L., & Brady, J. (2021). Together we can do so much: A systematic review and conceptual framework of collaboration in schools. *Canadian Journal of School Psychology, 36*(1), 59–85. https://doi.org/10.1177/0829573520915368

Gulley, A. P., Smith, L. A., Price, J. A., Prickett, L. C., & Ragland, M. F. (2017). Process-driven math: An auditory method of mathematics instruction and assessment for students who are blind or have low vision. *Journal of Visual Impairment & Blindness, 111*(5), 465–471.

Hall, G. E., & Hord, S. M. (2011). Implementation: Learning builds the bridge between research and practice. *Journal of Staff Development, 32*(4), 52–57.

Hallam, P. R., Smith, H. R., Hite, J. M., Hite, S. J., & Wilcox, B. R. (2015). Trust and collaboration in PLC teams: Teacher relationships, principal support, and collaborative benefits. *NASSP Bulletin, 99*, 193–216. https://doi.org/10.1177/0192636515602330

Hart, J. E., & Brehm, J. (2013). Promoting self-determination: A model for training elementary students to self-advocate for IEP accommodations. *Teaching Exceptional Children, 45*(5), 40–48. https://doi.org/10.1177/004005991304500505

Harris, M. (2019). *A future for planning: Taking responsibility for twenty-first century challenges*. Routledge.

Hawley, C. E., Cardoso, E., & McMahon, B. T. (2013). Adolescence to adulthood in STEM education and career development: The experience of students at the intersection of underrepresented minority status and disability. *Journal of Vocational Rehabilitation, 39*(3), 193–204. https://doi.org/10.3233/JVR-130655

Herzberg, T. S., & Rosenblum, L. P. (2014). Print to braille: Preparation and accuracy of mathematics materials in K–12 education. *Journal of Visual Impairment & Blindness, 108*(5), 355–367. https://doi.org/10.1177/0145482X1410800502

Holbrook, C. M. (2015) Renewing and refreshing knowledge base of the field of visual impairment: A call to action. *Journal of Visual Impairment & Blindness, 109*(2), 159–162. https://doi.org/10.1177/0145482X1510900213

Hourcade, J. J., & Bauwens, J. (2003). *Cooperative teaching: Rebuilding and sharing the schoolhouse*. Pro-Ed.

Individuals with Disabilities Education Improvement Act of 2004 (IDEA), Pub. L. No. 108-446, § 118, Stat. 2647 (2004).

Jackson, H. G., & Neel, R. S. (2006). Observing mathematics: Do students with EBD have access to standards-based mathematics instruction? *Education and Treatment of Children, 29*(4), 593–614.

Jitngernmadan, P., Stöger, B., Petz, A., & Miesenberger, K. (2017). Extraction methodology of implicit didactics in math schoolbooks for the blind. *Student Health Technology and Informatics. 242*, 843–850. https://doi.org/10.3233/978-1-61499-798-6-843

Johnson, E. S., Zheng, Y., Crawford, A. R., & Moylan, L. A. (2019). Developing an explicit instruction special education teacher observation rubric. *The Journal of Special Education, 53*(1), 28–40. https://doi.org/10.1177/0022466918796224

Koenig, A. J. & Holbrook, M. C. (2000). Ensuring high-quality instruction for students in braille literacy programs. *Journal of Visual Impairment & Blindness, 94*(11), 677–694. https://doi.org/10.1177/0145482X0009401102

Kurz, A., Elliott, S. N., Lemons, C. J., Zigmond, N., Kloo, A., & Kettler, R. J. (2014). Assessing opportunity-to-learn for students with disabilities in general and special education classes. *Assessment for Effective Intervention, 40*(1), 24–39. https://doi.org/10.1177/1534508414522685

Lalvani, P. (2012). Parents' participation in special education in the context of implicit educational ideologies and socioeconomic status. *Education and Training in Autism and Developmental Disabilities, 47*(4), 474–486.

Lambert, R. (2015). Constructing and resisting disability in mathematics classrooms: A case study exploring the impact of different pedagogies. *Educational Studies in Mathematics, 89*(1), 1–18. https://doi.org/10.1007/s10649-014-9587-6

Lambert, R. & Tan, P. (2017). Conceptualizations of students with and without disabilities as mathematical problem solvers in educational research: A critical review. *Education Sciences, 7*(2), 51. https://doi.org/10.3390/educsci7020051

Lee, F. L., Yeung, A. S., Tracey, D., & Barker, K. (2015). Inclusion of children with special needs in early childhood education: What teacher characteristics matter. *Topics in Early Childhood Special Education, 35*(2), 79–88. https://doi.org/10.1177/0271121414566014

Little, M. E., & Dieker, L. A. (2015). Intensifying instruction and interventions within multitiered systems of support. In B. Witzel (Ed.), *Bridging the gap between arithmetic & algebra* (pp. 157–166). Council for Exceptional Children.

Magiera, K., Smith, C., Zigmond, N., & Gebauer, K. (2005). Benefits of co-teaching in mathematics classes. *Teaching Exceptional Children, 37*(3), 20–4. https://doi.org/10.1177/004005990503700303

Maguvhe, M. (2015). Teaching science and mathematics to students with visual impairments: Reflections of a visually impaired technician. *African Journal of Disability, 4*(1), 1–6. https://doi.org/10.4102/ajod.v4i1.194

Nolan-Spohn, H. (2016). Increasing student involvement in IEPs. *Mid-Western Educational Researcher*, *28*(3), 7.

Olivos, E. M. (2009). Collaboration with Latino families: A critical perspective of home–school interactions. *Intervention in School and Clinic*, 45(2), 109–115. https://doi.org/10.1177/1053451209340220

Olson, J. M. (2003). *Special education and general education teacher attitudes toward inclusion* [Unpublished master's thesis]. University of Wisconsin–Stout.

Organisation for Economic Co-operation and Development [OECD]. (2023). *Mathematics performance (PISA).* OECD Data. Retrieved November 10, 2023, from https://data.oecd.org/pisa/mathematics-performance-pisa.htm

Palawat, M., & May, M. E. (2012). The impact of cultural diversity on special education provision in the United States. *Journal of the International Association of Special Education*, *13*(1), 58–63.

Pearl, C., Dieker, L. A., & Kirkpatrick, R. M. (2012). A five-year retrospective on the Arkansas Department of Education Co-Teaching Project. *Professional Development in Education*, *38*(4), 571–587. https://doi.org/10.1080/19415257.2012.668858

Ray, J. A., Pewitt-Kinder, J., & George, S. (2009). Partnering with families of children with special needs. *Young Children*, *64*(5), 16–22.

Reinholz, D. L., & Shah, N. (2018). Equity analytics: A methodological approach for quantifying participation patterns in mathematics classroom discourse. *Journal for Research in Mathematics Education*, *49*(2), 140–177. https://doi.org/10.5951/jresematheduc.49.2.0140

Rose, D., & Meyer, A. (2008). *A practical reader in universal design for learning.* Harvard Press.

Rottmann, M., Colling, P., Hack, T.-P., Chan, R., Hüger, F., Schlicht, P., & Gottschalk, H. (2020). Prediction error meta classification in semantic segmentation: Detection via aggregated dispersion measures of softmax probabilities. *2020 International Joint Conference on Neural Networks*, 1–9. https://doi.org/10.1109/IJCNN48605.2020.9206659

Rule A. C., Stefanich G. P., Boody, R. M., Peiffer B. (2011). Impact of adaptive materials on teachers and their students with visual impairments in secondary science and mathematics classes. *International Journal of Science Education*, *33*(6), 865–887. https://doi.org/10.1080/09500693.2010.506619

Salend, S. J. (2011). *Creating inclusive classrooms: Effective and reflective practices* (7th ed.). Pearson.

Scruggs, T. E., Mastropieri, M. A., & McDuffie, K. A. (2007). Co-teaching in inclusive classrooms: A meta-synthesis of qualitative research. *Exceptional Children, 73*(4), 392–416. https://doi.org/10.1177/001440290707300401

Smith, D. W., & Smothers, S. M. (2012). The role and characteristics of tactile graphics in secondary mathematics and science textbooks in braille. *Journal of Visual Impairment & Blindness, 106*(9), 543–554. https://doi.org/10.1177/0145482X1210600905

Stoianov, I., & Zorzi, M. (2012). Emergence of a "visual number sense" in hierarchical generative models. *Nature Neuroscience, 15*(2), 194–196. https://doi.org/10.1038/nn.2996

Solis, M., Vaughn, S., Swanson, E., & McCulley, L. (2012). Collaborative models of instruction: The empirical foundations of inclusion and co-teaching. *Psychology in the Schools, 49*(5), 498–510. https://doi.org/10.1002/pits.21606

Supalo, C. A., Isaacson, M. D., & Lombardi, M. V. (2014). Making hands-on science learning accessible for students who are blind or have low vision. *Journal of Chemical Education, 91*(2), 195–199. https://doi.org/10.1021/ed3000765

Sutton, H. (2017), Students with disabilities as likely to enter STEM fields as those without disabilities. *Disability Compliance for Higher Education, 22*(9), 9. https://doi.org/10.1002/dhe.30292

Tan, P., Lambert, R., Padilla, A., & Wieman, R. (2018). A disability studies in mathematics education review of intellectual disabilities: Directions for future inquiry and practice. *The Journal of Mathematical Behavior, 54*, 100672. https://doi.org/10.1016/j.jmathb.2018.09.001

Tannock, R. (2009). ADHD with anxiety disorders. In T. E. Brown (Ed.), *ADHD comorbidities: Handbook for ADHD complications in children and adults* (pp. 131–155). American Psychiatric Publishing, Inc.

Test, D. W., & Neale, M. (2004). Using the self-advocacy strategy to increase middle graders' IEP participation. *Journal of Behavioral Education, 13*(2), 135–145. https://doi.org/10.1023/B:JOBE.0000023660.21195.c2

Turnbull, A., Turnbull, R., Wehmeyer, M. L., & Shogren, K. A. (2013). *Exceptional lives: Special education in today's schools*. Pearson.

Tucker, V., & Schwartz, I. S. (2013). Parents' perspectives of collaboration with school professionals: Barriers and facilitators to successful partnerships in planning for students with ASD. *School Mental Health, 5*(1), 3–14. https://doi.org/10.1007/s12310-012-9102-0

U.S. Bureau of Labor Statistics (2023). *Top 10 fastest growing occupations* [Infographic]. https://www.bls.gov/emp/graphics/fastest-growing-occupations.htm

Villa, R. A., & Thousand, J. S. (Eds.). (2005). *Creating an inclusive school.* ASCD.

Wehmeyer, M. L. (2006). Beyond access: Ensuring progress in the general education curriculum for students with severe disabilities. *Research and Practice for Persons With Severe Disabilities, 31*(4), 322–326. https://doi.org/10.1177/154079690603100405

Wehmeyer, M. L., Palmer, S. B., Shogren, K., Williams-Diehm, K., & Soukup, J. H. (2013). Establishing a causal relationship between intervention to promote self-determination and enhanced student self-determination. *The Journal of Special Education, 46*(4), 195–210. https://doi.org/10.1177/0022466910392377

Zhou, X., Wei, W., Zhang, Y., Cui, J., & Chen, C. (2015). Visual perception can account for the close relation between numerosity processing and computational fluency. *Frontiers in Psychology, 6,* 1364. https://doi.org/10.3389/fpsyg.2015.01364

APPENDIX

ADDITIONAL RESOURCES

The following list of resources is also available in digital format at the official website for the American Printing House for the Blind. Visit https://www.aph.org/aph-press/aph-press-learning for an accessible version of this list.

ABACUS

Abacus for Beginners
An hour-long introduction to the abacus from APH
https://www.youtube.com/watch?v=2ExtjevMp1Q

Abacus Video Trainings
Videos from Iowa Educational Services for the Blind and Visually Impaired
https://www.iesbvi.org/educator-resources/math-2

Abacus YouTube Videos
Video resources from the Texas School for the Blind
https://www.youtube.com/@VideoTSBVI/search?query=abacus

The Abacus: A Million Manipulatives in Your Pocket
Online course from APH Hive for teaching and adapting problems using the abacus
https://aphhive.org/#/courses/enroll/2265 (free login required)

Cranmer Abacus
Allows students to perform computations quickly and independently
https://www.aph.org/product/cranmer-abacus

Millaway, S. M. (2025). *Abacus basic competency: A counting method*. American Printing House for the Blind.
New publication, available in large print and BRF
https://www.aph.org/product/abacus-basic-competency-a-counting-method-large-print

Poppe, K. J. (1997). *Beginner's abacus: Instructions for use*. American Printing House for the Blind.
Available for free on APH site
https://sites.aph.org/files/manuals/7-03180-00.pdf

Practice2Master Abacus App
Free iOS and Android app for interactive abacus learning
https://www.aph.org/product/practice2master-abacus

Sewell, D. & Rose, J. (2019). T*he counting method for the Cranmer abacus*. Texas School for the Blind and Visually Impaired.
Available in print and digital formats

BRAILLE (NEMETH CODE)

Braille Authority of North America (2022). *The Nemeth braille code for mathematics and science notation*, 2022. American Printing House for the Blind.
Comprehensive resource for transcribers, proofreaders, and software developers
https://www.brailleauthority.org/sites/default/files/2025-01/Nemeth_2022_3.pdf

Cleveland, J., Bean, J., Bird, M., Kelley, S., O'Brien, S., Osterhaus, S., Sewell, D., & Torrence, G. (2017). ***Nemeth at a glance: A math resource, grade-level chart, and evaluation tool.*** **Texas School for the Blind and Visually Impaired.**
Available in print and digital formats

Feel 'n Peel Stickers: Nemeth Braille–Print Numbers 0–100
Multiuse tactile stickers
https://www.aph.org/product/feel-n-peel-stickers-nemeth-braille-print-numbers-0-100

Flip-Over Concept Books: Fractions
Books that teach students about fractions, decimals, and percentages
https://www.aph.org/product/flip-over-concept-book-fractions-nemeth

Learning and Teaching the Nemeth Code Within UEB Contexts: A Step-by-Step Guide
Available in print, digital, and braille formats
https://www.pathstoliteracy.org/resource/learning-and-teaching-the-nemeth-code

Math Drill Cards: Number and Math Sign Cards (Nemeth)
Flash cards that reinforce math skills
https://www.aph.org/product/math-drill-cards-number-and-math-sign-cards-nemeth

Mathematics Braille Transcribing
Free Nemeth code transcription course from the National Federation of the Blind
https://nfb.org/programs-services/braille-certification/mathematics-braille-transcribing

The Nemeth Braille Code Curriculum
Includes a PreK–second grade curriculum, focused lessons, and a

Nemeth symbol library
https://www.pathstoliteracy.org/nemeth-curriculum

Nemeth Code Curriculum: A Trusted Resource for Teaching Nemeth Code
A curriculum developed by the National Federation of the Blind in partnership with the Pearson Accessibility Team for Assessments
https://nfb.org/programs-services/education/
nemeth-code-curriculum

Nemeth Focused Lessons
Paths to Literacy helps students increase knowledge and understanding of key mathematical concepts
https://www.pathstoliteracy.org/nemeth-braille-code-focused-lessons

Nemeth in a Box for Middle School Students
Paths to Literacy has lessons for teachers to introduce and review Nemeth code
https://www.pathstoliteracy.org/
nemeth-box-middle-school-students

Nemeth Online Tutorial
From the Texas School for the Blind and Visually Impaired and Paths to Literacy
https://www.pathstoliteracy.org/resource/
nemeth-online-tutorial-tsbvi

Nemeth Reference Sheets
Downloadable in PDF and BRF formats
https://www.pathstoliteracy.org/resource/nemeth-reference-sheets

Nemeth Tutorial
APH teaches students and teachers the many levels of Nemeth code
https://nemeth.aphtech.org

Project Inspire, Course 3
Intro to mathematical Nemeth code symbols used in Grades 2–5
https://www.pathstoliteracy.org/nemeth-code-symbols-used-grades-2-5-course-3

BRAILLE (UEB MATH/SCIENCE)

Flip-Over Concept Books: Fractions
Books that teach students about fractions, decimals, and percentages
https://www.aph.org/product/flip-over-concept-book-fractions-ueb

The Hitchhiker's Guide to UEB Mathematics
Designed as a quick resource and memory jogger
https://www.banzat.org.nz/documents/HHGMaths.pdf

ICEB Guidelines for Technical Material, 2014
Comprehensive guidelines from the creators of UEB, the International Council on English Braille
https://www.iceb.org/ueb.html

Math Drill Cards: Number and Math Sign Cards (UEB)
Flash cards that reinforce math skills
https://www.aph.org/product/math-drill-cards-number-math-signs-ueb

UEB Math Tutorial
APH teaches students and teachers the many levels of UEB Math
https://uebmath.aphtech.org

UEB Reference Sheets for Math
Available in large print and braille from National Braille Press
https://shop.nbp.org/products/ueb-reference-sheets-for-math

Unified English Braille Codebooks and Guidance Documents
Available in PDF and BRF formats from the Braille Authority of North America
https://www.brailleauthority.org/unified-english-braille-codebooks

CLASSROOM RESOURCES

All-in-One Board
Multiplatform, versatile, and easily adjustable for classroom use
https://www.aph.org/product/all-in-one-board

Classroom Posters of the Mathematical Practices
Available for various grade bands
https://www.debbiewaggoner.com/math-practice-standards.html

Consumable Hundreds Chart
Includes blank forms for customization
https://www.aph.org/product/consumable-hundreds-chart

Flip-Over Concept Books: Line Paths
Randomize and search for adjacent panels involving line paths
https://www.aph.org/product/flip-over-concept-books-line-paths

Flip-Over Concept Books: Textures
Randomize and search for adjacent panels involving textures
https://www.aph.org/product/flip-over-concept-books-textures

Math Symbol Reference Booklets
Large-print resource for students and teachers using Nemeth and UEB math
https://www.aph.org/product/math-symbol-reference-booklets-large-print-version

Math Window
Utilizes magnetic tiles that combine braille and print for lessons in mathematics
https://www.mathwindow.com

Number Line Device
Helps students who are blind and visually impaired comprehend abstract numerical concepts
https://www.aph.org/product/number-line-device

Student Mathematical Practices Posters
Downloadable bundle for Grades 6–8
https://www.amsti.org/math-6-8-classroom

Tactile Book Builder Kit
Build a book and create a connection to literacy
https://www.aph.org/product/tactile-book-builder-kit

Tactile Connections: Symbols for Communication
Kit for constructing an augmentative communication system for learners who are nonverbal and who have multiple disabilities
https://www.aph.org/product/tactile-connections-symbols-for-communication

Tactile Town: 3-D O&M Kit
An interactive 3-D model that teaches spatial concepts and cognitive mapping skills
https://www.aph.org/product/tactile-town-3-d-om-kit

Tactile Treasures Kit: Math and Language Concepts for Young Children With Visual Impairments, Tactile–Color Edition
Informal assessment and teaching tool
https://www.aph.org/product/tactile-treasures-kit-tactile-color-edition

30 Fantastic Free Fourth Grade Math Games
From We Are Teachers
https://www.weareteachers.com/fourth-grade-math-games

Trifold Board
Versatile hook–loop board
https://www.aph.org/product/trifold-board

FRACTIONS

Cuisenaire Rods
Classic manipulatives for learning a variety of mathematical concepts, including fractions
https://www.cuisenaire.co.uk

MathBuilders: Fractions
Manipulatives for young students to learn fractions, mixed numbers, and decimals in both UEB and Nemeth code
https://www.aph.org/product/mathbuilders-fractions

Practice2Master Fractions App
Free and accessible iOS app for helping students become proficient in fraction calculations
https://www.pathstoliteracy.org/resource/practice2master-fractions-app

Teaching Fractions With Appropriate Materials
Resources for making tactile fraction manipulatives
https://www.pathstoliteracy.org/teaching-fractions-with-appropriate-materials

Vision Impaired Fractions
TSVI-created fractions curricula

https://www.teacherspayteachers.com/browse?search=vision%20 impaired%20fractions

GEOMETRY

Desmos Geometry Tool
Online geometry drawing
https://www.desmos.com/geometry

Geometro Shapes
3-D manipulative shapes to aid in teaching geometry
https://www.aph.org/product/geometro-gs22-medium-set

Parallel Lines
Video resource from the Texas School for the Blind and Visually Impaired
https://library.tsbvi.edu/Play/15373

Perpendicular Lines
Video resource from the Texas School for the Blind and Visually Impaired
https://library.tsbvi.edu/Play/15344

Project Inspire, Course 4
Intro to geometry and tactile graphics for students Grades 3–8
https://www.pathstoliteracy.org/project-inspire-geometry-and-tactile-graphics-students-grades-3-8-course-4

Skew Lines
Video resource from the Texas School for the Blind and Visually Impaired
https://library.tsbvi.edu/Play/16472

GRAPHING TOOLS

Bold Line Tactile Graph Sheets
Tool to help teachers and transcribers create charts and graphs for students who are blind and visually impaired
https://www.aph.org/product/bold-line-tactile-graph-sheets

Embossed Graph Sheets
Available in 0.5-inch, 0.75-inch, and 1-inch versions
https://www.aph.org/product/low-relief-embossed-graph-sheets-0-5-inch-squares-20-x-20-grid
https://www.aph.org/product/embossed-graph-sheets-0-75-inch-squares
https://www.aph.org/product/embossed-graph-sheets-1-inch-squares-2

Graph Benders
Create durable graphs and tactile diagrams and charts
https://www.aph.org/product/graph-benders

Graphic Aid for Mathematics
Create figures and graphs related to arithmetic, algebra, geometry, trigonometry, and calculus
https://www.aph.org/product/graphic-aid-for-mathematics-2

MathBuilders: Statistics
Designed for young blind and low vision math learners in Grades K–3
https://www.aph.org/product/mathbuilders-statistics/

SAS Graphics Accelerator
Accessible software for creating, exploring, and sharing data visualizations
https://support.sas.com/software/products/graphics-accelerator/index.html

GRAPHING CALCULATORS AND DEVICES

Cadence Tablet
Pocket-sized multiline electronic braille reader
https://www.tactile-engineering.com/cadence

Desmos Graphing Calculator
Online graphing calculator software
https://www.desmos.com/calculator

Graphiti
Tablet braille reader compatible with the Orion TI-84 Plus Talking Graphing Calculator
https://www.orbitresearch.com/product/graphiti

Monarch
APH's 10-line-by-32-cell refreshable braille display; includes access to APH's Tactile Graphics Image Library
https://www.aph.org/meet-monarch

Orion TI-84 Plus Talking Graphing Calculator
Accessible version of the Texas Instruments classic
https://www.orbitresearch.com/product/orion-ti-84-plus

MEASUREMENT

Analog Clock Model
For demonstration of how an analog clock works
https://www.aph.org/product/analog-clock-model

Braille–Large Print Protractor
Pivoting wand allows for tactile exploration of angles
https://www.aph.org/product/braille-large-print-protractor

Classroom Calendar
Large print and braille calendars for all students
https://www.aph.org/search-results/?fwp_search_term=classroom%20calendar

Clock Face Sheets in Braille
Tactile sheets each containing four embossed braille clock faces
https://www.aph.org/product/clock-face-sheets-in-braille

Draw2Measure Protractor App
Free iOS alternative for measuring angles
https://www.aph.org/product/draw2measure-protractor-app-for-ios-devices-free-app

iBill
Free currency reader from the U.S. Bureau of Engraving and Printing
https://www.bep.gov/services/currency-accessibility/us-currency-reader-program

Individual Calendar Kit
Students can create their own large print and braille calendars
https://www.aph.org/product/individual-calendar-kit

MathBuilders: Measurement
Designed for young blind and low vision math learners in Grades K–3
https://www.aph.org/product/mathbuilders-measurement

Ruler: 1-Foot Braille English Measurement
Includes a caliper slide to aid in accuracy
https://www.aph.org/product/ruler-1-foot-braille-metric-english-measurement

Tactile Caliper
Provides a braille output to a sixteenth of an inch
https://shop.nbp.org/products/tactile-caliper

Tactile Demonstration Thermometer
Available in UEB and Nemeth code
https://www.aph.org/product/tactile-demonstration-thermometer-ueb-nemeth

Tactile Protractor
Braille protractor for classroom and practical use
https://www.aph.org/product/tactile-protractor-4-pack

MATH MANIPULATIVES

Base-10 Blocks
10 flat manipulative blocks per set
https://www.aph.org/product/focus-in-mathematics-base-ten-blocks-flats-10-per-set

Brannan Cubarithm Slate
A rubber frame and grid with braille cubes that can be used to set up mathematical problems in braille
https://www.aph.org/product/brannan-cubarithm-slate

Consumable Number Lines
Blank number lines for instructors and students to customize
https://www.aph.org/product/consumable-number-lines

Desktop Stick-On Number Lines
Reusable visual and tactile number lines
https://www.aph.org/product/desktop-stick-on-number-lines-large-print-braille-5-pack

Digi-Blocks
Nested math manipulatives
https://louisekool.com/collections/digi-blocks

Hundreds Board and Manipulatives
Set includes numbers and shapes
https://www.aph.org/product/hundreds-board-manipulatives-nemeth

Interlocking Base Ten Starter Set
Helps students understand abstract base-10 concepts
https://www.learningresources.com/item-interlocking-plastic-base-ten-starter-set

Magna-Tiles
Magnetic building sets for all ages
https://www.magnatiles.com

Number Lines from APH
A variety of number line products
https://www.aph.org/search-results/?fwp_search_term=number+line

Place Value Setter
Designed for Grade 1–3 students as they learn to represent large numbers using place value and written digits
https://www.aph.org/product/place-value-setter

Omnifix Cubes
Manipulative tools for exploring spatial relationships
https://www.didax.com/omnifix-cubes-set-of-100.html

StackUps Kit: Spatial Reasoning Using Cubes and Isometric Drawings
Teaches students to interpret raised-line graphics depicting 3-D

figures
https://www.aph.org/product/stackups-kit-spatial-reasoning-using-cubes-and-isometric-drawings

Tactile Algebra Tiles
Magnetic tiles that can be manipulated on a folding board to assist students in solving math problems
https://www.aph.org/product/tactile-algebra-tile

Tactile Five and Ten Frames
Supports counting and cardinality, composing and decomposing numbers, place value, and addition and subtraction skills
https://www.aph.org/product/tactile-five-and-ten-frames

Tactile Tangrams Kit
Teaches the relationships between parts and wholes
https://www.aph.org/product/tactile-tangrams-kit

Unifix Cubes
Manipulatives also available in virtual versions
https://www.didax.com/unifix-cubes.html

Wikki Stix
Colorful, bendable craft and activity sticks made of wax and yarn
https://www.wikkistix.com

Work–Play Trays
Enclosed spaces in which students can play, explore, and learn independently
https://www.aph.org/search-results/?fwp_search_term=work%20play%20tray

ONLINE COURSES AND PROFESSIONAL DEVELOPMENT FOR TEACHERS

APH Hive Brick-by-Brick Numeracy
Class that addresses adapting math games and activities using LEGO Braille Bricks
https://aphhive.org/#/courses/enroll/2131 (free but login required)

APH Hive Tactile Aids for Mathematics
Explore a variety of adapted mathematical tools and their use
https://aphhive.org/#/courses/enroll/7 (free but login required)

Project Inspire 2: Access and Equity in STEM Learning for Individuals Who Read Braille
A total of 10 free classes for professionals and middle and high school students
https://www.pathstoliteracy.org/resource/project-inspire

How to Implement the UDL Guidelines for Mathematics
Lesson tips from Novak Education
https://info.novakeducation.com/udl-guidelines-math

National Council of Teachers of Mathematics Principles to Actions Toolkit for Teachers
Grade-band-specific modules
https://www.nctm.org/PtAToolkit

Quorum Project
Data science professional development course
https://quorumlanguage.com

Standards for Mathematical Practice: Commentary and Elaborations for K–5
Annotations and guidelines for elementary grades

https://mathematicalmusings.org/wp-content/uploads/2014/02/Elaborations.pdf

Standards for Mathematical Practice: Commentary and Elaborations for 6–8
Annotations and guidelines for middle grades
https://mathematicalmusings.org/wp-content/uploads/2014/05/2014-05-06-Elaborations-6-8.pdf

UDL Lesson Plan Makeovers
Math lesson improvement tips
https://www.novakeducation.com/blog/udl-lesson-plan-makeovers-math-lessons-math-resource-download

ONLINE RESOURCES

Digi-Blocks
Videos of mathematics program for all students in Grades Pre–6 and for remediation, intervention, and special education students of all ages
https://www.youtube.com/user/digiblock

IXL Math
Interactive talking math practice
https://www.ixl.com/math

Khan Academy
Free tutorials on all subjects, including math
https://www.khanacademy.org

MAPs, PATH, and Circle of Friends
Resources for inclusionary practices
https://www.inclusion.com/path-maps-and-person-centered-planning

Mathematics Made Easy for Children With Visual Impairment
PDF of the international edition
https://icevi.org/wp-content/uploads/2019/05/Mathematics-Made-Easy-for-Children-with-Visual-Impairment.pdf

NASA Data Sonifications
Explore the universe through sound
https://www.nasa.gov/data-sonifications

National Center on Intensive Intervention at the American Institute of Research
Lesson plans and tips for instructors teaching counting skills
https://intensiveintervention.org

Project Math Access
A series of video lessons on teaching accessible mathematical concepts from numeracy to algebra
https://www.youtube.com/@projectmathaccess2562

64 Oz. Games
Accessibility kits for popular board games
https://www.64ouncegames.com

Teaching and Learning Mathematics With the Common Core
Video series by NCTM and The Hunt Institute
https://hunt-institute.org/resources/2015/04/nctm-mathematics-video-series-standards-for-mathematical-practice

Tinkercad
Open-source 3-D coding and printing
https://www.tinkercad.com

TACTILE GRAPHICS

Aluminum Diagramming Foil
Two-sided sheets for creating tactile graphics masters and drawing objects
https://www.aph.org/product/aluminum-diagramming-foil-sheets

BANA Net Example
Sample tactile graphic designs
https://brailleauthority.org/tg/web-manual/u6nets.html

BANA's Guidelines and Standards for Tactile Graphics
Available online and in PDF and BRF files
https://www.brailleauthority.org/guidelines-and-standards-tactile-graphics

Draftsman Tactile Drawing Board
Create instant raised-line drawings
https://www.aph.org/product/draftsman-tactile-drawing-board

Picture Maker Wheatley Tactile Diagramming Kit
Create maps, charts, shapes, and other tactile diagrams
https://www.aph.org/product/picture-maker-wheatley-tactile-diagramming-kit

Quick-Draw Paper
Works with water-based markers to create tactile lines and pictures
https://www.aph.org/product/quick-draw-paper

Tactile Graphics Kit
An extensive set of tools and materials that allows instructors to create custom raised-line maps, graphs, diagrams, and charts
https://www.aph.org/product/tactile-graphics-kit

TactileDoodle
Create instantly raised graphics and pictures in combination with tactile drawing film and a stylus or ballpoint pen
https://www.aph.org/product/tactiledoodle-kit

Tactiles
Open-source 3-D printing resources
https://www.tactiles.eu

TactiPad
Includes mathematical tools for this raised-line drawing board
https://thinkable.nl/tactipad

INDEX

Page numbers in *italics* refer to figures.

www.ingramcontent.com/pod-product-compliance
Ingram Content Group UK Ltd.
Pitfield, Milton Keynes, MK11 3LW, UK
UKHW051453290926
15471UKWH00008B/1181

9 781950 723263